Datacenter Design and Management

A Computer Architect's Perspective

Synthesis Lectures on Computer Architecture

Synthesis Lectures on Computer Architecture publishes 50- to 100-page publications on topics pertaining to the science and art of designing, analyzing, selecting and interconnecting hardware components to create computers that meet functional, performance and cost goals. The scope will largely follow the purview of premier computer architecture conferences, such as ISCA, HPCA, MICRO, and ASPLOS.

Datacenter Design and Management: A Computer Architect's Perspective
Benjamin C. Lee
2016

A Primer on Compression in the Memory Hierarchy
Somayeh Sardashti, Angelos Arelakis, Per Stenström, and David A. Wood
2015

Research Infrastructures for Hardware Accelerators
Yakun Sophia Shao and David Brooks
2015

Analyzing Analytics
Rajesh Bordawekar, Bob Blainey, and Ruchir Puri
2015

Customizable Computing
Yu-Ting Chen, Jason Cong, Michael Gill, Glenn Reinman, and Bingjun Xiao
2015

Die-stacking Architecture
Yuan Xie and Jishen Zhao
2015

Single-Instruction Multiple-Data Execution
Christopher J. Hughes and
2015

Power-Efficient Computer Architectures: Recent Advances
Magnus Själander, Margaret Martonosi, and Stefanos Kaxiras
2014

FPGA-Accelerated Simulation of Computer Systems
Hari Angepat, Derek Chiou, Eric S. Chung, and James C. Hoe
2014

A Primer on Hardware Prefetching
Babak Falsafi and Thomas F. Wenisch
2014

On-Chip Photonic Interconnects: A Computer Architect's Perspective
Christopher J. Nitta, Matthew K. Farrens, and Venkatesh Akella
2013

Optimization and Mathematical Modeling in Computer Architecture
Tony Nowatzki, Michael Ferris, Karthikeyan Sankaralingam, Cristian Estan, Nilay Vaish, and David Wood
2013

Security Basics for Computer Architects
Ruby B. Lee
2013

The Datacenter as a Computer: An Introduction to the Design of Warehouse-Scale Machines, Second edition
Luiz André Barroso, Jimmy Clidaras, and Urs Hölzle
2013

Shared-Memory Synchronization
Michael L. Scott
2013

Resilient Architecture Design for Voltage Variation
Vijay Janapa Reddi and Meeta Sharma Gupta
2013

Multithreading Architecture
Mario Nemirovsky and Dean M. Tullsen
2013

Performance Analysis and Tuning for General Purpose Graphics Processing Units (GPGPU)
Hyesoon Kim, Richard Vuduc, Sara Baghsorkhi, Jee Choi, and Wen-mei Hwu
2012

Automatic Parallelization: An Overview of Fundamental Compiler Techniques
Samuel P. Midkiff
2012

Phase Change Memory: From Devices to Systems
Moinuddin K. Qureshi, Sudhanva Gurumurthi, and Bipin Rajendran
2011

Multi-Core Cache Hierarchies
Rajeev Balasubramonian, Norman P. Jouppi, and Naveen Muralimanohar
2011

A Primer on Memory Consistency and Cache Coherence
Daniel J. Sorin, Mark D. Hill, and David A. Wood
2011

Dynamic Binary Modification: Tools, Techniques, and Applications
Kim Hazelwood
2011

Quantum Computing for Computer Architects, Second Edition
Tzvetan S. Metodi, Arvin I. Faruque, and Frederic T. Chong
2011

High Performance Datacenter Networks: Architectures, Algorithms, and Opportunities
Dennis Abts and John Kim
2011

Processor Microarchitecture: An Implementation Perspective
Antonio González, Fernando Latorre, and Grigorios Magklis
2010

Transactional Memory, 2nd edition
Tim Harris, James Larus, and Ravi Rajwar
2010

Computer Architecture Performance Evaluation Methods
Lieven Eeckhout
2010

Introduction to Reconfigurable Supercomputing
Marco Lanzagorta, Stephen Bique, and Robert Rosenberg
2009

On-Chip Networks
Natalie Enright Jerger and Li-Shiuan Peh
2009

The Memory System: You Can't Avoid It, You Can't Ignore It, You Can't Fake It
Bruce Jacob
2009

Fault Tolerant Computer Architecture
Daniel J. Sorin
2009

The Datacenter as a Computer: An Introduction to the Design of Warehouse-Scale Machines
Luiz André Barroso and Urs Hölzle
2009

Computer Architecture Techniques for Power-Efficiency
Stefanos Kaxiras and Margaret Martonosi
2008

Chip Multiprocessor Architecture: Techniques to Improve Throughput and Latency
Kunle Olukotun, Lance Hammond, and James Laudon
2007

Transactional Memory
James R. Larus and Ravi Rajwar
2006

Quantum Computing for Computer Architects
Tzvetan S. Metodi and Frederic T. Chong
2006

Datacenter Design and Management: A Computer Architect's Perspective
Benjamin C. Lee

ISBN: 978-3-031-00624-1 paperback
ISBN: 978-3-031-01752-0 ebook

DOI 10.1007/978-3-031-01752-2

A Publication in the Springer series
SYNTHESIS LECTURES ON ADVANCES IN AUTOMOTIVE TECHNOLOGY

Lecture #37
Series Editor: Margaret Martonosi, *Princeton University*
Series ISSN
Print 1935-3235 Electronic 1935-3243

Datacenter Design and Management

A Computer Architect's Perspective

Benjamin C. Lee
Duke University

SYNTHESIS LECTURES ON COMPUTER ARCHITECTURE #37

ABSTRACT

An era of big data demands datacenters, which house the computing infrastructure that translates raw data into valuable information. This book defines datacenters broadly, as large distributed systems that perform parallel computation for diverse users. These systems exist in multiple forms—private and public—and are built at multiple scales. Datacenter design and management is multifaceted, requiring the simultaneous pursuit of multiple objectives. Performance, efficiency, and fairness are first-order design and management objectives, which can each be viewed from several perspectives. This book surveys datacenter research from a computer architect's perspective, addressing challenges in applications, design, management, server simulation, and system simulation. This perspective complements the rich bodies of work in datacenters as a warehouse-scale system, which study the implications for infrastructure that encloses computing equipment, and in datacenters as distributed systems, which employ abstract details in processor and memory subsystems. This book is written for first- or second-year graduate students in computer architecture and may be helpful for those in computer systems. The goal of this book is to prepare computer architects for datacenter-oriented research by describing prevalent perspectives and the state-of-the-art.

KEYWORDS

computer organization and design, energy efficiency, cluster computing, data centers, distributed systems, cloud computing, performance evaluation methodologies, resource allocation, software scheduling

Contents

Preface

This synthesis lecture is written for first- or second-year graduate students in computer architecture. The reader is expected to have completed graduate coursework in computer architecture; additionally, a course in distributed systems would be helpful. Moreover, the reader is expected to have some basic knowledge and prior experience in using the tools of the trade: cycle-level simulators. This background provides the requisite perspective on benchmarking and simulation for conventional workloads to help the reader appreciate challenges that are new and unique to datacenter workloads.

Moreover, this lecture may be helpful for graduate students in computer systems. Because the determinants of datacenter performance and efficiency increasingly lie at the hardware-software interface, architecture and systems perspectives on datacenters could be integrated to reveal new research directions. Because datacenter operators are rightfully wary of introducing new hardware into well-tuned systems, architects must anticipate system management challenges during architectural design. Furthermore, emerging hardware technologies and architectures require new system organizations and management.

Related Lectures. This synthesis lecture complements two existing synthesis lectures, one on datacenters and another on performance evaluation methods. Both lectures are highly recommended for their breadth and complementary perspective. Barroso et al. present a lecture on datacenters that focuses on the design of warehouse-scale machines, which is often taken to mean the datacenter itself [16]. For example, the lecture describes the facility, the peripheral infrastructure that supports the computing equipment, and figures of merit for evaluating datacenter efficiency and costs (e.g., total cost of ownership). This lecture is highly recommended for its breadth, its focus on warehouse-scale systems and facilities, and its industry-strength perspective. In contrast, our lecture focuses on processor and memory design, and emphasizes experimental methodologies that draw on a rich body of widely deployed open-source applications.

Eeckhout's lecture on performance evaluation methods focuses on performance evaluation methodologies, with a specific emphasis on strategies that accelerate the evaluation process [42]. The lecture describes analytical performance models that concisely represent processor performance. It also describes varied statistical strategies that reveal application performance while reducing the number of instructions simulated with cycle-level timing models. These research methodologies are best suited for understanding broad, general-purpose benchmark suites. In contrast, we focus on datacenter workloads and full system simulation. Previously proposed strategies for rapid design space exploration may apply to datacenter workloads as well, but we would need to adapt them to full system simulation.

We organize this lecture on datacenter research methodologies in several chapters. Chapter 2 describes several representative datacenter applications, surveys their implications for hardware architectures, and proposes benchmarking strategies. Chapters 3–4 survey recent research in server design and management. Chapter 5 details strategies for simulating datacenter servers. Specifically, we present approaches to processor and memory simulation, and demonstrate methodologies for precise simulations that target application regions of interest. Finally, Chapter 6 describes strategies for simulating datacenter dynamics at scale. We present analytical and empirical approaches to understanding task behaviors and queueing dynamics.

Collectively, the goal of this book is to prepare a computer architect for datacenter-oriented research. It describes prevalent perspectives and the state-of-the-art. Yet, for all the research that is surveyed in this book, many challenges remain and the required advances in datacenter design and management would very much benefit from a computer architect's perspective.

Benjamin C. Lee
January 2016

Acknowledgments

This synthesis lecture draws from varied research projects that have spanned the last eight years. During this time, I have been fortunate to work with extraordinary collaborators and students. My perspectives on computer architecture and datacenter systems have been enriched by these collaborations. I am thankful for my collaborators during projects at Microsoft Research—Vijay Janapa Reddi, Trishul Chilimbi, and Kushagra Vaid—an institution where fundamental research is leavened with applied perspectives. I am also grateful for my collaborators during projects at Stanford University—Mark Horowitz, Christos Kozyrakis, and Krishna Malladi—a place where interdisciplinary work is pervasive. Finally, I am incredibly fortunate to work with a group of outstanding Ph.D. students at Duke University—Songchun Fan, Marisabel Guevara, Ziqiang Huang, Tamara Lehman, Bryan Prosser, Qiuyun Wang, Seyed Majid Zahedi, and Pengfei Zheng.

For this project, I am thankful for Margaret Martonosi and Michael Morgan's invitation to write a synthesis lecture on datacenter research. The lecture has benefited substantially from reviews by Christina Delimitrou, Stijn Eyerman, Boris Grot, and Lingjia Tang. In advance, I thank the faculty and students who use and share this lecture to advance their own research projects. I would very much appreciate any feedback from researchers and practitioners.

Lastly, I would like to thank our research sponsors for their support of my projects and the students who contribute to them. Our projects have been supported by the National Science Foundation, Duke University, and the Google Faculty Research Award. They have also been supported by STARnet, a Semiconductor Research Corporation program, sponsored by MARCO and DARPA. Any opinions, findings, or recommendations expressed in this material are those of the author and do not necessarily reflect the views of these sponsors.

Benjamin C. Lee
January 2016

CHAPTER 1
Introduction

An era of big data demands datacenters, which house the computing infrastructure that translates raw data into valuable information. Data flows are increasingly diverse and multi-directional. Early Internet services supplied data, which users translated into information. In contrast, today's services supply, consume, and analyze data to produce information that is timely and suited to each user's individual context. The increasing sophistication of data sources and flows motivate correspondingly capable datacenters.

1.1 DATACENTERS DEFINED

This book defines datacenters broadly as large distributed systems that perform parallel computation for diverse users. Defined as such, datacenters exist in multiple forms. Private datacenters compute for users within a single organization. Within an organization, sharing increases utilization and improves efficiency by amortizing system costs over more work. Yet sharing requires mechanisms to distribute resources according to policies for fairness and priority, which are tailored to organizational needs. Internet services (e.g., search, mail) and software-as-a-service are representative of this type of datacenter computing.

In contrast, public datacenters compute for users that share no affiliation with the system operator. Users select machines from a menu and pay as these resources are reserved or consumed. This market for computation requires users to characterize their workloads and request resources explicitly. Infrastructure-as-a-service is representative of this type of datacenter computing.

Datacenters exist at varied scales. Industry's largest datacenters incorporate thousands of machines that compute for millions of tasks, yet smaller datacenters are far more common. This book is agnostic about scale, emphasizing the similarities between highly parallel and distributed clusters rather than the unique characteristics of warehouse-scale clusters. For example, recent research in search engines consider diverse document corpora, from the world wide web to online encyclopedia. At all scales, datacenters encounter fundamental challenges.

1.2 RESEARCH DIRECTIONS

Datacenter design and management is multifaceted, requiring the simultaneous pursuit of multiple objectives. First, system architects balance service quality and energy efficiency. Computational tasks must complete within tens of milliseconds for online services and should progress at the same rate as others in the job to avoid stragglers that lengthen the critical path. Yet the system should colocate software to amortize hardware power, employ low-power servers when possible, and power only the servers required to meet service targets. Such efficiency would increase datacenter capability within the same footprint.

System architects must also balance service quality and fairness. The definition of fairness depends on the system context and is expressed in terms of equality in allocation or progress, in terms of domain-specific priorities, or in terms of game-theoretic desiderata. In cooperative settings, such as data- and task-parallel computation, equal progress reduces the variance in latency distribution and guards against outliers. In competitive settings, such as shared clusters and federated datacenters, economic notions of fairness incentivize sharing and guard against envy. Fairness, or the lack thereof, determines whether strategic users choose to participate in a shared system.

The pursuit of system desiderata requires advances in management mechanisms. Datacenter profiling must supply the requisite data for intelligent allocation and scheduling. Datacenter allocation must distribute physical hardware such as server, processor, or memory resources to diverse tasks. It must also distribute less tangible but no less critical resources such as time, priority, and power. Each of these resources are made scarce by design within efficient datacenters, making constrained management and optimization a rich vein of research.

1.3 RESEARCH CHALLENGES

Computer architects who wish to perform datacenter research encounter five broad challenges—applications, design, management, server simulation, system simulation—all of which demand strategies and methodologies that extend beyond their conventional tools.

Applications and Benchmarks. First, datacenter applications are increasingly built atop generalizable platforms and frameworks that implement multiple abstraction layers. The abstractions are intended to separate the programming model with which users develop distributed algorithms and the run-time system that breaks computation into many small, short tasks. MapReduce is an example of this strategy in practice—the programming model defines Map and Reduce functions and the run-time system creates individual Map and Reduce tasks that compute on small slices of the data. The abstractions are also intended to increase modularity, which permits sophisticated libraries for stream processing, machine learning, and graph analytics to leverage capabilities from lower in the stack of system software.

In this setting, computer architects must clearly define the workload of interest. They cannot simply benchmark an application because benchmarking datacenter applications is replete with subtle questions. They should clarify whether benchmarks evaluate the run-time system or specific types of tasks. Should a Spark benchmark include both the engine and the application tasks? What index of documents and set of queries produce a representative benchmark for search? To answer these questions, computer architects require increasingly precise measurements from increasingly large software systems.

Design. Second, hardware design is motivated by two competing objectives in datacenter computing—service quality and energy efficiency. Today's hardware architectures are designed for one or the other. High-performance design ensures low latency and high throughput accompanied by significant power costs whereas low-power design ensures energy efficiency accompanied by significant performance risks. During design, architects characterize hardware-software interactions to explore a rich space of high-performance and low-power design points. The selection of multiple, heterogeneous designs from this space dictates the balance between performance and efficiency. Moreover, the organization of these heterogeneous components at scale affects the competition and contention for preferred resources.

Management. Third, datacenter architects must design for manageability, which means anticipating the run-time consequences of decisions made at design-time. Datacenters co-locate multiple software tasks on a single hardware platform to amortize its fixed power costs, a strategy that improves efficiency and energy proportionality. At run-time, the datacenter must allocate hardware and schedule software to meet performance targets and ensure fairness. Architects consider a combination of throughput and latency when evaluating performance; the former describes system capacity whereas the latter describes application responsivenes. Furthermore, architects must adopt definitions of fairness that encourage users to employ shared datacenter hardware instead of procuring private systems. Studies in resource allocation might take a game-theoretic approach to account for strategic behavior. Because server and datacenter designs vary in management complexity, architects should optimize designs to manage performance risk—the probability of contention and poor outcomes—in shared systems.

Hardware Simulation. Fourth, hardware simulation for datacenter applications is complicated by the system software stack. Datacenter applications require operating systems, virtual machines, and libraries. Moreover, they perform network and disk I/O. For example, the open source engine for web search is implemented in Java and requires the Java virtual machine. To support the full system software stack and the hardware ecosystem, computer architects must rely on a sophisticated combination of emulation, which ensures functional correctness, and simulation, which provides timing models for greater insight.

Computer architects must apply application-specific insight to identify regions of interest for full system simulation with detailed timing models, which is prohibitively expensive under normal system operation. For tractability, systems and applications should initialize and warm up in emulation mode before performing specific tasks of interest in simulation mode. Application tasks are short and getting shorter, requiring only tens or hundreds of milliseconds of computation. If architects could start simulation right before the task of interest, simulating a task to completion would be tractable.

System Simulation. Finally, system simulation for datacenter applications is complicated by system scale. Datacenter applications increasingly rely on run-time systems to schedule and distribute tasks to workers. Hardware performance interacts with queueing dynamics to determine distributions for task response time. Datacenter operators often optimize latency percentiles (e.g., 99th percentile) and mitigate stragglers in the distribution's long tail.

Computer architects can turn to queueing models for rules of thumb and back-of-the-envelope calculations to assess system throughput, queueing times, and waiting times. When certain assumptions are satisfied, M/M/1 queueing models provide an elegant analytical framework for reasoning about system dynamics. But under arbitrary and general settings with non-parametric distributions for task inter-arrival and service times, architects should rely on discrete event simulation. Both analytical and empirical approaches can play a role when assessing system performance at scale, a setting in which physical measurements on a deployed system can be impractical.

CHAPTER 2

Applications and Benchmarks

Datacenter applications share a number of characteristics. First, they compute on big data, which motivates new architectures that blend the capacity and durability of storage with the performance of memory. Increasingly, data resides in distributed memory and is accessed via the network according to Zipfian popularity distributions [14]. Such distributions have long tails such that the most popular pieces of data are accessed far more frequently than less popular pieces. Although Zipfian distributions' temporal locality facilitates memory caching, their long tails produce irregular requests and demand high-capacity memory.

Second, datacenter applications extend to warehouse-scale with task parallelism. Data analysis is partitioned into many small pieces to form tasks, which the system queues and schedules. Third, applications are implemented atop programming models and run-time systems, such as MapReduce, to expose and manage task parallelism. Distributed computing frameworks implement abstractions such that programmers need not reason explicitly about a computing cluster's physical implementation when specifying program functionality. Collectively, these application characteristics have aided the proliferation of distributed computing.

2.1 BENCHMARK SUITES

As interest in datacenter research grows, benchmark suites have proliferated. A number of studies survey datacenter workloads, providing computer architects multiple perspectives on benchmarking. Many of these workloads are open source and widely available. The difficulty is exercising them with realistic input data and computational kernels, and measuring their hardware activity precisely.

Benchmark suites identify representative applications and provide software targets for research in hardware systems. Datacenter benchmarks are distinguished by their fine-grained tasks that facilitate "scale-out" computation in large distributed systems. In contrast, conventional benchmarks often focus on "scale-up" computation, which benefits from single-threaded performance. Scale-out and scale-up systems require different benchmarks and demand different design decisions. During design space exploration, for example, scale-out workloads might trade fewer resources (e.g., smaller cores and caches) for the power efficiency that is required for tightly integrated datacenter servers.

Lim et al. study representative datacenter workloads and design server architectures that balance performance, power, and cost [101]. The benchmark suite includes search, mail, content serving, and map reduce. As scale-out workloads, these benchmarks exhibit fine-grained task parallelism, which demands less capability from any one hardware component. Lim et al. find that datacenter workloads can benefit from low-end and mobile-class hardware, which offer performance with attractive total-cost-of-ownership (TCO). This seminal study illustrates new metrics and methodologies for the design of server architectures.

Ferdman et al. present CloudSuite, a benchmark suite for scale-out workloads. The suite includes data serving, MapReduce, media streaming, SAT solver, web frontend, and web search [49]. The difficulty with datacenter workloads is not obtaining the software but deploying it on realistic hardware platforms and exercising it with representative tasks. CloudSuite documents its system settings and makes several observations about microarchitectural activity—datacenter workloads miss often in the instruction cache, exhibit little instruction- or memory-level parallelism, work with data that does not fit in on-chip caches, and require little inter-core communication. The CloudSuite infrastructure and its insights enable a number of server design studies [47, 48, 56, 104].

Wang et al. present BigDataBench with data stores, relational databases, search engines, and graph analysis [159]. BigDataBench is notable for broad coverage of applications and data inputs, which affect microarchitectural activity and pose methodological challenges for cycle-level simulation for computation on realistic datasets. BigDataBench documents its system settings and makes several observations about system behavior—datacenter workloads exhibit low application intensity as defined by the number of instructions per byte of data transferred from memory, miss often in the instruction cache, and benefit from larger last-level caches. Characterizing and identifying a modest number of representative workloads and data inputs from the complete set improves tractability [75, 76].

In this chapter, we survey four representative datacenter applications—web search, memory caching and storage, MapReduce, and graph analysis. Search and caching represent latency-sensitive applications in which computation must complete in tens of milliseconds to guarantee the user experience. MapReduce and graph analysis represent batch applications in which computation may complete in the background. We consider these applications from a computer architect's perspective, describing the computation and its demands on the system architecture.

2.2 SEARCH

Data volumes are growing exponentially, driven by a plethora of platforms for content creation, from web pages and blogs to news feeds and social media. Users, confronted with this massive data corpus, turn to search engines for tractability. These engines identify and deliver relevant data in response to a user's query. Search engines have become increasingly sophisticated, partly to keep pace with data diversity. For example, early web search engines used the PageRank algorithm to identify popular pages that are often cited and linked by other popular pages [24]. Today's web search combines PageRank with many other algorithms that use contents from pages and queries from users to determine relevance [55]. Future search engines may consider hundreds of features, which describe multiple types of content and users, and employ statistical machine learning to estimate relevance from this high-dimensional space [55, 137, 138].

Search is a natural starting point for research in datacenter architecture. Search is a high profile application that needs little motivation for the broader research community. Moreover, it embodies several fundamental characteristics of datacenter applications. First, search is relevant at multiple scales. Although Google search and Microsoft Bing dominate mind share for web search in the United States, researchers can study realistic search engines at more experimentally accessible scales by indexing and querying data on a personal machine, a subset of the Internet (e.g., Wikipedia), or any other well-defined data corpus of interest. Second, search is parallelizable. A strategy that distributes indexed data across multiple machines and then sends a query to each of those machines scales well on commodity hardware. In this setting, a deep analysis of one search engine node generalizes to others in the datacenter. Finally, search is extensible. Atop a search engine's primitive functionality—quantifying relevance given a query specification—researchers can explore variants that anticipate user questions by translating simple queries into more sophisticated ones or estimate relevance with novel machine learning methods.

Search Engine. The search engine performs two types of computation. First, the engine crawls and indexes content with batch computation that aggregates and organizes data. Crawls may be performed at many scales—although Google search crawls the world wide web, Wikipedia search crawls only the entries in its online encyclopedia. The engine creates an index from crawled pages, which contains information about words and their locations in the data corpus. Google describes its search index much like an index in the back of a book; a reader locates the search term within the index to find the appropriate page(s) [55]. An index may become quite large, especially if it spans the entire web. The engine typically partitions an index such that each part fits in a server's main memory. In a multi-server system, each query is distributed across servers and accesses the index in parallel.

Second, the search engine executes interactive queries, identifying related pages with the index and calculating relevance scores to determine the pages to display. The engine calculates a static score for each page, independently of any query, based on attributes such as popularity (e.g., PageRank) and freshness. Upon receiving a query, the engine accounts for query terms and operators to calculate a dynamic score. The engine orders pages according to relevance scores, prepares captions or snippets for the most relevant pages, and serves these pages to the user. In a multi-server system, each server sends the most relevant pages in its index to an aggregator, which produces the final set of results.

Figure 2.1 illustrates the organization of the Microsoft Bing Search engine [137, 138]. Queries enter the system via the top-level aggregator, which caches results for popular or repeated queries [154]. The aggregator distributes queries to index-serving nodes, which calculate dynamic page ranks (using a neural network in this example) for their respective parts of the index. Finally, the aggregator integrates nodes' results to serve links and page snippets to users.

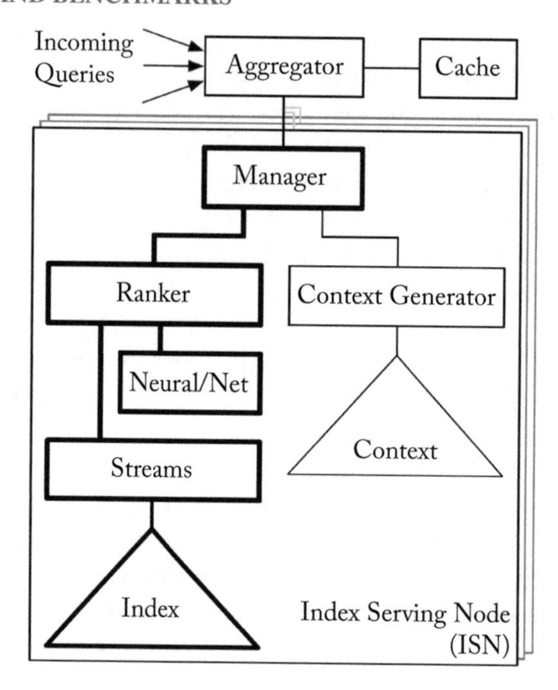

Figure 2.1: Overview of Microsoft Bing and the scoring computation within an index serving node [137, 138].

Query Complexity. A search engine must serve relevant content for heterogeneous and diverse queries. Figure 2.2 illustrates the space of possible query constructions. A query in this space is defined by the number of terms, the logical operators that integrate them into a single query, and the wildcard operators that allow a query to express rich permutations of any one term. Many queries execute only one term and do not use operators. Yet the majority of queries are more sophisticated.

The frequency distribution of query complexity is well studied. Silverstein et al. performed an early study using Altavista logs from 1999 [147] and Table 2.1 summarizes their findings. More than 55% of queries specify multiple terms. And 20% of queries use one or more operators to connect these terms together. Empty queries contain no terms and usually result from technical misunderstandings. For example, users fail to enter terms into an advanced search interface, which supports the use of query operators.

These early observations about user behavior and query complexity have proven durable. Jansen et al. performed a later study in 2006 to show that query lengths have increased over time as users and search engines have become more sophisticated [74]. Indeed, search engines themselves increasingly introduce complexity on the user's behalf. For example, the engine may translate a user search for "rabbit" into a broader search that includes variants (e.g., "rabbits") and synonyms (e.g., "hare," "bunny") for more comprehensive and relevant results [41].

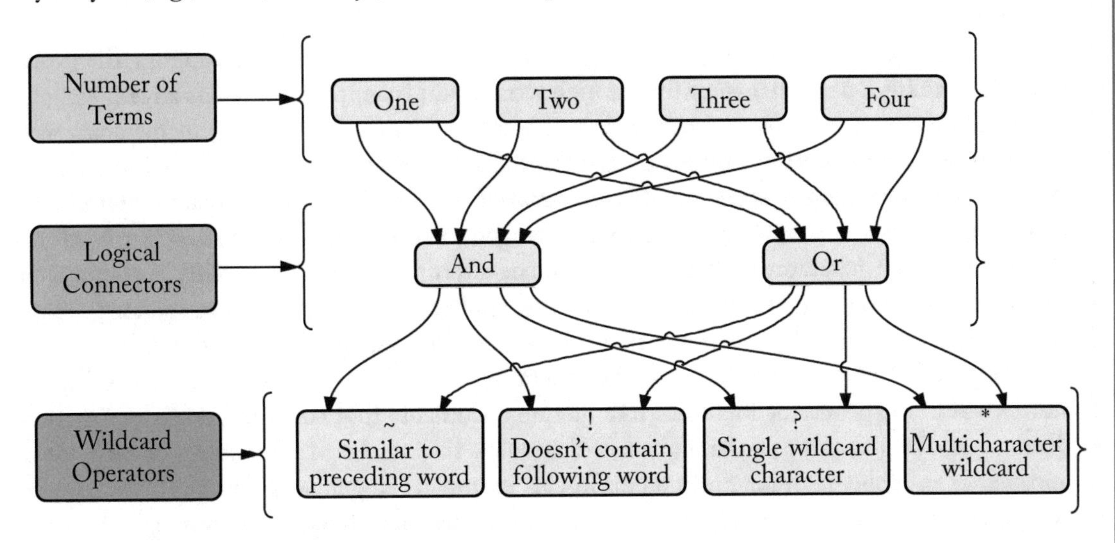

Figure 2.2: Determinants of query complexity.

Table 2.1: Distribution of query complexity [147]

No. of Terms (%)	No. of Operators (%)
0 (20.5%)	0 (79.6%)
1 (25.8%)	1 (9.7%)
2 (26.0%)	2 (6.0%)
3 (15.0%)	3 (2.6%)
>3 (12.6%)	>3 (2.1%)

A study of web search must account for query complexity, exercising the search engine with queries that combine terms and operators in ways that users might. We describe a query generator that produces diverse queries for a search engine benchmark [23]. Inputs to the generator are frequency distributions on query length and operator count, such as the one in Table 2.1, and indexed documents that provide a dictionary from which words are sampled. Outputs are queries that reflect the complexity of user behavior and the popularity of content within a data corpus.

Specifically, the generator creates a dictionary of words using contents for indexed documents. The dictionary tracks the frequency of each word's appearance in these documents. To produce a single-term query, the generator randomly samples a word from the dictionary. The probability of sampling a particular word is proportional to its popularity. However, single-term queries do not reflect the sophistication of modern searches. Even in 1999, more than half of all queries included two or more terms. And this trend toward multi-term queries has increased over the years [74, 176].

To produce a multi-term query, the generator performs conditional sampling. The probability of sampling a particular word for the n-th term depends on the $n - 1$ terms already sampled for the same query; such dependences reflect true user behavior. Specifically, the generator produces a multi-term query by sampling a starter word from indexed documents. Then the generator examines the documents that contain this starter word. Within these documents, the starter word appears in context with other, related words. The generator creates a set of neighboring words. From this set, the generator randomly selects the next term. Selection probability is proportional to each word's frequency in the set. By recursively applying this process, the generator produces a query with the desired length and with a coherent mix of semantically related words.

Figure 2.3 illustrates the process for generating a multi-term query, using "rabbit" as the starter word. The generator identifies 632 unique documents that contain "rabbit." With these documents, the generator constructs a sub-dictionary that tracks the frequency of each word's appearance in rabbit-relevant documents. In this example, "Roger" and "framed" are the most typical words that appear with "rabbit." The generator performs weighted sampling on words in the sub-dictionary to produce the second word in the query. Suppose the second word is "lucky." The query is now comprised of the two words "rabbit" and "lucky." If a third word is needed, the generator identifies documents that match both terms to create a new sub-dictionary for sampling. Thus, the generator recursively adds terms to produce a query with the desired length. The query length itself is drawn from a probability distribution, such as the one in Table 2.1.

As query length increases, the number of relevant documents decreases sharply. For example, consider the frequency distribution of words within Wikipedia documents. The vast majority of words in these documents appear less than ten times. A much smaller number of words are common and appear in more than 45K of the 50K Wikipedia documents. Such a distribution has two related implications. During query generation, the size of sub-dictionaries decreases rapidly with query length. And during query processing, the number of results returned to the user decreases rapidly with query complexity.

Logical operators connect words in a multi-term query. The most common operators are AND and OR. Of these, AND is the default. These operators affect the number of documents that match a given query. Using the AND operator sharply reduces the number of documents returned to a user. However, returned documents are often more relevant since the AND operator prunes those that only contain one of the query's terms. Another class of operators, wildcards, increases query sophistication and complexity. Table 2.2 summarizes these operators. For example, the NEAR operator ($\sim$) finds words that are similar to the user-specified word. Similarity is defined by the Levenshtein distance—all words that differ in one character from the original word are also considered relevant (e.g., "text" and "test"). Wildcard operators increase processor intensity by broadening query scope. Although users rarely specify explicit wildcards, these operators

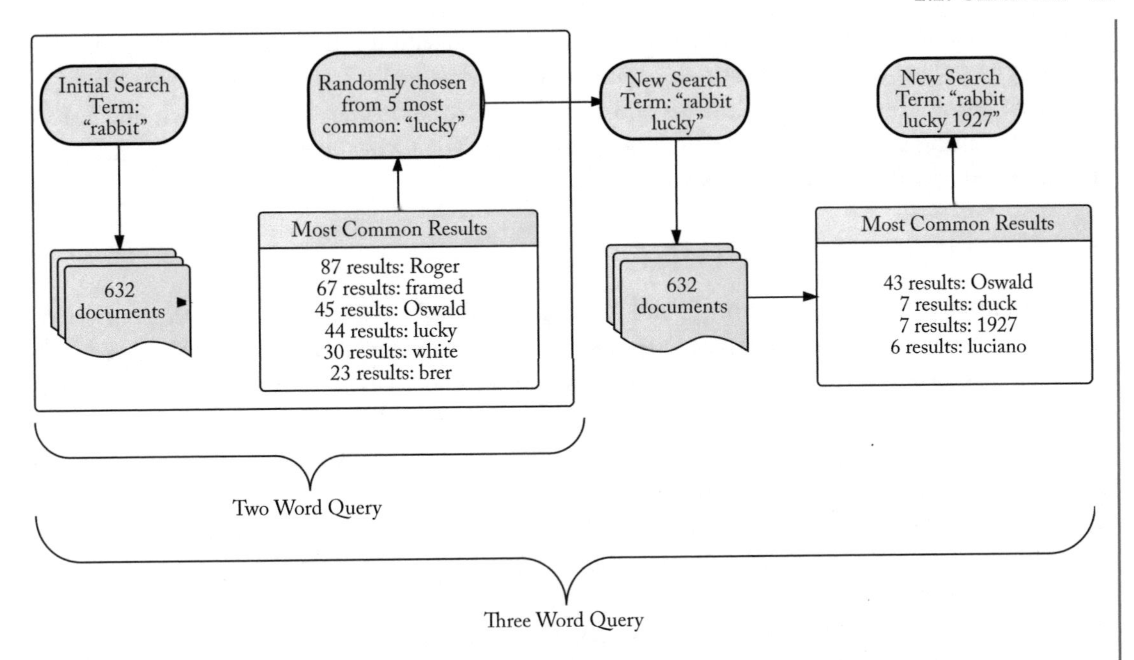

Figure 2.3: Example for multi-term query generation.

can be inserted transparently by the search engine to improve service quality. For example, the NEAR operator can be used to mitigate spelling errors and provide other user-friendly features.

Table 2.2: Wildcard operators with usage examples

Op	No. of Terms (%)	No. of Operators (%)
?	single-character wildcard	"thing?" matches "thing," "things," etc.
*	multi-character wildcard	"a*t" matches "at," "apt," "assistant," etc.
!	inverts following word/phrase	"!orange" returns any doc without "orange"
~	finds words within specied edit distance from preceding word	"text" matches "text," "test," "next," etc.

Benchmarking Strategies. Benchmarking web search for computer architecture research requires analysis in several dimensions—the data corpus, the queries that exercise the search engine, and the hardware design space. In several regards, web search is easy to simulate. First, the index is partitioned across many servers, which rank pages in parallel, and simulating a single server provides insights across all servers. Second, the index is sized to fit in memory, which reduces or eliminates the network and storage I/O that is not often accommodated by cycle-level simulators. Finally, simulating end-to-end query execution is tractable because an individual query completes in tens or hundreds of milliseconds. Practically, researchers have simulated web search queries by using MARSSx86 and DRAMSim2, deploying Apache Nutch and Lucene, and executing search queries for Wikipedia's data corpus. Other methodologies for full system simulation may be equally effective.

2.3 MEMORY CACHING

Conventional approaches to storage and retrieval fail to scale as data volumes increase. Disk data transfer rates, which have failed to keep pace with disk capacity, limit how frequently each data block can be accessed, on average. Ousterhout et al. find that the ratio of disk to bandwidth capacity constrains the system, permitting access to a given 1 KB data block only once every 58 days [122]. Such poor performance, which is attributed to disk latency, raises the possibility of stranding data in the storage system.

The root cause of this performance challenge is poorly matched disk capacity and bandwidth. System architects might address this imbalance by distributing the same amount of data across many more disks, thereby increasing system bandwidth. Because data volume remains constant as the number of disks increases, disk utilization falls and the cost per bit rises. As disk-based storage becomes more expensive, today's DRAM and Flash memories become fast and viable alternatives. Such economic trends would also open the door to emerging technologies, such as phase change memory.

Table 2.3: Evolution in disk technology over 25 years. Reproduced from Ousterhout et al. [122].

	Mid-1980s	2009	Difference
Disk capacity	30 MB	500 GB	16,667 better
Maximum transfer rate	2 MB/sec	100 MB/s	50× better
Latency (seek + rotate)	20ms	10 nms	2× better
Capacity/bandwidth (large blocks)	15 s	5,000 s	333× worse
Capacity/bandwidth (1KB blocks)	600 s	58 days	8,333× worse

Datacenter servers with large main memories reduce access latency by caching popular data. Heavy-tailed Zipfian distributions on data popularity means that memory caches can significantly improve performance at moderate cost. At warehouse-scale, caching frameworks pool many servers' memories into a single, large, logical cache accessed via a key-value system. Industry uses a variety of key-value storage systems. Indeed, Facebook caches 75% of its non-media data in memory [115, 122].

Key-Value Storage Systems. We survey recent advances in key-value storage systems. Strikingly, many of these system software advances respond to hardware and technology trends. FAWN is a key-value store designed for low-power processors and Flash memory. RAMCloud is a key-value store designed to ensure durability and availability given DRAM and low-latency networks. And because the network is on the critical path, bypassing the kernel or accelerating the network protocol with hardware support is compelling. Fundamentally, these research directions provide foundations for fine-grained, low-latency access to data at warehouse-scale.

A key-value storage system provides fine-grained access to large data sets. In each of these systems, data is represented as a collection of key-value pairs; each value is associated with a unique key. A hash function maps a key to the corresponding value's location, which identifies the server node and a physical location in memory or storage. The hash tables support a simple set of operations—`get(k)` retrieves a data value with key `k`, `set(k,v)` stores a data value `v` with key `k`, and `delete(k)` deletes a data value with key `k`. Atop these primitives, systems researchers can build sophisticated data models for consistency, reliability, etc. Internet service providers deploy such systems at scale—Amazon uses Dynamo [36], LinkedIn uses Voldemort [120], and Facebook uses memcached [118].

The **memcached** framework is a system for distributed memory caching [118]. Data requests issue queries to memcached before querying back-end databases. Consistent hashing directs each query to a memcached node by mapping a key to a unique server with the corresponding data value. Within each server, memcached divides memory into slabs, each of which are designated to hold data objects of a particular size. Slabs and their data sizes are chosen to reduce fragmentation, increase utilization, and increase cache hit rates [115].

Facebook's memcached deployment is dominated by `get` queries, serving thirty `get`'s for every `set` [14]. This balance suggests that its applications treat distributed memory as a persistent store rather than a temporary cache. The majority of its stored values are less than 500 bytes each and nearly all stored values are less than 1,000 bytes. Memcached values are often found in the cache and hit rates are high—greater than 90%.

Performance optimizations for distributed memory caching require a holistic approach that couples software data structures with hardware capabilities. Lim et al. design **MICA** (memory-store with intelligent concurrent access) to deliver high performance for in-memory, key-value stores [98]. First, MICA partitions data to exploit multi-core parallelism within a backend server. Second, it interfaces directly with the network interface card and avoids latency overheads in the operating system's network stack. Finally, it uses novel data structures for memory allocation and indexing.

Li et al. further pursue performance by coupling MICA's software advances with balanced server architectures [97]. The principle of balanced system design in which resources for compute, memory, and network are provisioned in the right proportions for the workload improves efficiency, reduces latency, and increases throughput to a billion requests per second on a single node. Moreover, modern platforms provide new capabilities, such as direct cache access, multi-queue network interface cards, and prefetching, which enhance request throughput when used properly.

Pursuing performance and durability, Ousterhout et al. propose **RAMCloud**, a key-value store that holds all data in memory and obviates the need for hard disks [122]. The case for

RAMCloud is based on software trends and hardware limitations. First, datacenters already hold a large majority of their data in memory and the marginal cost of holding the remainder in memory is no longer prohibitive. Second, datacenter applications can no longer tolerate disk's performance limitations—when disk is several orders of magnitude slower than memory even rare misses in the memory cache will have a large impact on average access time.

To implement the vision of placing all data in memory, RAMCloud addresses two challenges. For durability, it backs up data to disk. It distributes backups to hundreds or thousands of disks, which provide the requisite transfer rates for fast recovery [117]. For performance, RAMCloud must reduce network latency. DRAM latency ranges from tens to hundreds of nanoseconds. In contrast, Ethernet and the TCP/IP stack require hundreds of microseconds for round-trip transmission. Much of this time is spent in interrupt handling and protocol processing within the operating system. TCP/IP offload and kernel bypass can reduce round-trip software overheads to one microsecond [122].

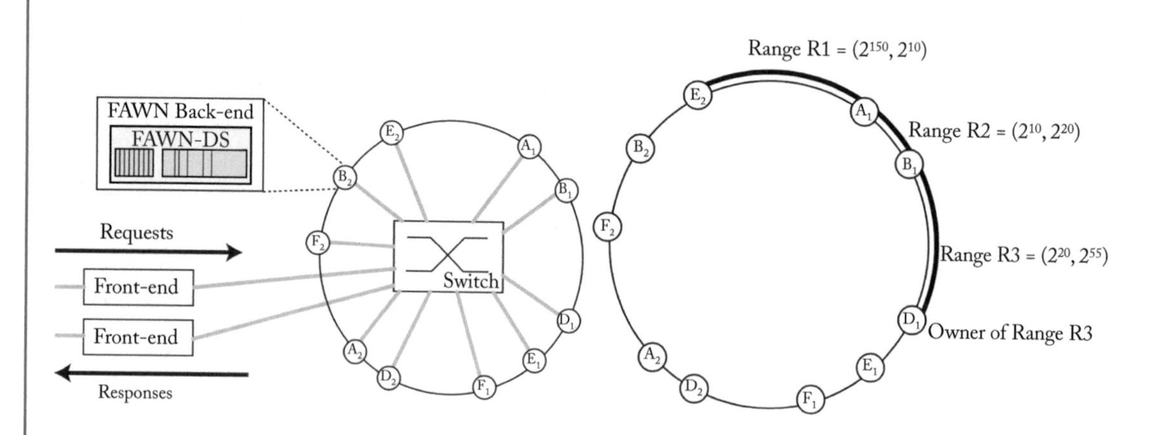

Figure 2.4: FAWN, a fast array of wimpy nodes, and its key-value store. Reproduced from Andersen et al. [2].

Pursuing energy efficiency for key-value stores, Andersen et al. propose **FAWN**, a key-value system designed for a fast array of wimpy nodes that employ low-power processors and solid-state disks [2, 3]. Compared to systems with general-purpose processors and hard disks, FAWN increases the number of queries per Joule by an order of magnitude or more for small, random accesses.

FAWN is comprised of a key-value system (KV) and a datastore (DS). The KV distributes data across datacenter nodes by mapping key ranges to DS nodes using consistent hashing. In this hashing scheme, a ring represents the continuum of key values. Each DS in the system is placed on the ring and its position determines the range of keys for which it is responsible. The KV routes each request, which uses a get/put interface, to the DS responsible for its key. For example, requests for keys between 2^{10} and 2^{20} are all routed to the same DS. With consistent hashing, only a fraction of the keys are re-mapped when DS nodes join or leave the system.

The DS serves each request, reading data from and writing data to solid-state disks. The DS is a log-structured store, which appends writes to the end of a data log. Appends are particularly well suited for solid-state disks because they avoid the overheads of garbage collection and coarse-grained erasure. To read data from the log, the DS maintains an in-memory hash table that maps a key in its assigned range to the corresponding location. This location is specified as an offset to the append-only log.

Benchmarking Strategies. Benchmarking distributed memory caching frameworks requires a sequence of queries to the system (e.g., get, set, and delete). Moreover, it requires a distribution for data popularity, which specifies how often each key is used within a key-value store. Finally, the data object size matters and, in memcached, the typical size is small. Most objects are KBs in size and objects are rarely larger than 200 MBs [14, 100]. Small, fine-grained accesses illustrate the memory caching's advantages over disk. But they also highlight the network's latency overheads.

The design of recent key-value stores and the servers that deploy them suggest that further reductions in access latency require a holistic strategy that accelerates processing for hash index lookups, reduces network overheads, and exploits emerging memory technologies. Further research is needed to determine whether targeted microbenchmarks that measure separate progress toward each of these goals is possible. For example, network interfaces that bypass the operating system may be orthogonal to advances in resistive memory technologies. Collectively, recent performance optimizations have produced multiple, independent advances toward more responsive access to big data.

2.4 MAPREDUCE

Big data is cumbersome, especially when many small pieces of computation are needed on every piece of data. Datacenters must distribute computation across tens of thousands of servers, orchestrate communication between those servers, and ensure performance and reliability at scale. These challenges appear in varied settings, from counting words to indexing web pages. Software developers ought to share the programming models and infrastructure that address these challenges rather than re-architect their own solutions. In such a setting, Dean and Ghemawat devised MapReduce [35].

MapReduce is both a programming model and a run-time system. The model expresses computation as a series of map and reduce functions, which compute on key-value pairs. The system performs computation by partitioning input data, launching map and reduce tasks at scale, and ensuring resilience with mechanisms to mitigate straggling or failed tasks. By separating the programming model and the run-time system, MapReduce provides clean abstractions. Software developers specify only map and reduce functions, knowing that the run-time system will manage their deployment in the datacenter. Indeed, these abstractions have fostered a large ecosystem of workloads built atop MapReduce for distributed databases, machine learning, graph analytics, and many others.

Programming Model. The MapReduce programming model is defined by its two constituent computational phases. Map applies a function to each element of the input data to produce a series of key-value pairs. Reduce applies a function to intermediate results in these key-value pairs to produce an integrated result. A MapReduce programmer specifies each of these functions and data types for key-value pairs to effect the desired computation.

```
map(key, value):
   // key: document name
   // value: document contents
   for each word w in value:
      EmitKeyValue(w, 1);
```

Figure 2.5: Map definition for word count, modified from [35].

```
reduce(keys, values):
   // keys: a list of words
   // values: a list of counts
   for each unique k in keys
      int result = 0;
      for each v in values with key k:
          result += v;
      Emit(k, result)
```

Figure 2.6: Reduce definition for word count, modified from [35].

Figures 2.5–2.6 reproduce an example from Dean and Barroso that counts words in a document. The map function consumes a key-value pair, which specifies data sources, and produces another pair, which indicates the occurrence of a word. The reduce function consumes these many key-value pairs and sums these results for each unique key. This example illustrates MapReduce in its simplest form, highlighting distinct map and reduce phases. These primitives extend naturally to varied settings, especially when many partial answers can be aggregated with commutative and associative reduce functions.

System Implementation. Implementations of MapReduce systems are notable for their scalability and resilience, both of which are derived from the nature of Map and Reduce primitives. The MapReduce programming model was inspired by primitives in functional programming languages, which treat functions as the fundamental building block for programs and treat computation as the composed evaluation of those functions. Purely functional programming languages do not produce side effects or modify global state. In this spirit, the MapReduce framework makes no provision for functions with side effects, shifting the burden of ensuring atomicity and consistency to the programmer. In practice, applications that produce side-effects are a poor match for the MapReduce model.

The lack of side effects reduces the difficulty of parallelizing map and reduce functions for distributed computation in a datacenter. A MapReduce implementation includes a master, which orchestrates data movement, and workers that perform either map or reduce computation. MapReduce splits input data into pieces and distributes data to map workers. These workers compute intermediate key-value results, store them in local disk, and communicate their location to the master. The master communicates these locations to the reduce workers. These workers read the key-value pairs, sort them by key, and apply the reduce function.

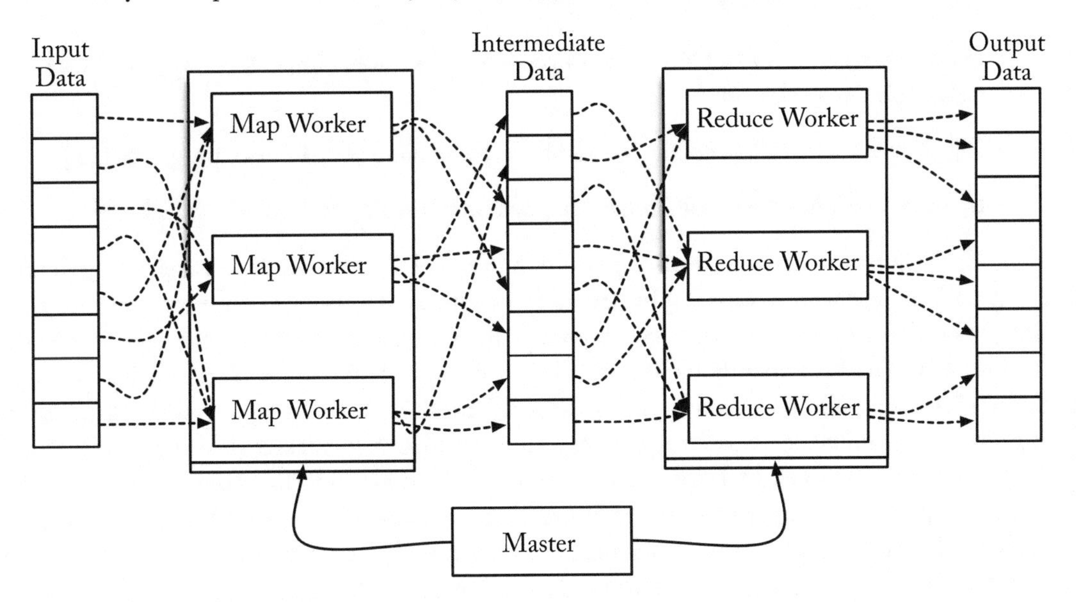

Figure 2.7: MapReduce overview.

By separating the MapReduce programming model from its implementation with a clean abstraction layer, system architects can pursue designs that ensure performance and resilience. For performance, MapReduce identifies straggling tasks and preemptively re-schedules them on another, potentially faster, machine. MapReduce tasks retrieve data from a distributed file system that replicates data and distributes them on many machines, motivating sophisticated task placement strategies that pursue locality. These same system properties ensure resilience. The master resets and reschedules failed tasks, and the distributed file system replicates data to guard against loss.

Variants of MapReduce. Since Google's introduction of MapReduce as a programming model for parallel computing in 2004, the systems community has produced a number of implementations. In 2007, Microsoft Research proposed Dryad for distributed data analysis on large clusters [73]. Dryad supports sophisticated communication topologies, expressed as directed acyclic graphs, that subsume MapReduce's restricted map-shuffle-reduce communication pattern. As in MapReduce, Dryad's implementation includes a run-time system that schedules tasks and manages resources.

In 2011, after six years of development, the Apache Software Foundation released Hadoop. Hadoop includes a distributed a file system (HDFS) and an implementation of MapReduce that mirrors Google's system software. Since its launch, Hadoop has provided a foundation for an ecosystem of software libraries and frameworks:

- **Hive** provides a database engine that transforms SQL-like queries into MapReduce tasks executed against HDFS [8]
- **HBase** provides a non-relational storage system that holds big data in sparse tables [7], as in Google's BigTable [29]
- **Pig** provides a scripting language to write sophisticated MapReduce programs [12]
- **Spark** and **Mahout** provide libraries for machine learning atop Hadoop [10, 13]

Spark is a particularly noteworthy addition to the Hadoop ecosystem, accelerating iterative computation for machine learning by orders of magnitude by using main memory more effectively. MapReduce and its variants communicate in an acyclic data flow model that, in practice, requires expensive I/O operations to a distributed file system. Such I/O is highly inefficient for the iterative computation found in many statistical machine learning applications (e.g., logistic regression). For efficiency, Spark introduces resilient distributed datasets (RDDs), a collection of data objects that can be cached in main memory across computational iterations. From a computer architect's perspective, Spark and other future advances present new opportunities. As software frameworks use hardware more efficiently, performance and energy efficiency constraints will shift from datacenter storage and networks to server processors and memory.

Finally, although we have focused our discussion of MapReduce on datacenter-scale implementations, the programming model can be implemented on shared memory multiprocessors as well [136, 153]. At such scale, the run-time system creates threads and distributes them across multiple cores. MapReduce's advantage is programmability and offers performance within competitive range of POSIX threads. Shared memory MapReduce implementations, which may become increasingly important with big memory servers, illustrate the value of clean abstraction layers between the programming model and its implementation.

Benchmarking Strategies. Benchmarking MapReduce for computer architecture research requires a data set and a sequence of Map and Reduce tasks. These tasks might arise from the simple applications that are described in Dean and Ghemawat's original study [35]—counting and sorting specific elements in a large data set. Alternatively, these tasks might arise from a sophisticated software framework built atop MapReduce for database queries or machine learning.

For example, the Spark framework supports a variety of operations on big data, which combine to implement broadly relevant statistical analyses such as logistic regression and alternating least squares.

Within sophisticated software frameworks built atop MapReduce, we can identify specific regions of interest for cycle-level simulation of processors and memory systems. Researchers have successfully simulated Phoenix MapReduce, an implementation for shared memory multiprocessors, and Spark MapReduce, a machine learning framework. At present, simulations model all hardware activity from the master, the worker(s), and other aspects of the run-time system. In the future, more precise measurements may target specific map or reduce tasks, and exclude run-time overheads.

2.5 GRAPH ANALYSIS

Datacenter workloads are more diverse than the applications and frameworks we have described thus far. For example, as frameworks for graph analysis mature, they may present specific demands for hardware. Graphs from datacenter computation tend to be large and sparse. Media companies (e.g., Netflix) may implement collaborative filtering to analyze users' movie preferences and tailor recommendations based on each user's viewing history. Social networking frameworks may implement page rank algorithms to analyze network structure and identify popular individuals or organizations in the graph. Several software frameworks provide clean programming abstractions and capable libraries for such computation.

Programming Models and Run-Time Systems. Low et al. devise the **GraphLab** framework for data mining and machine learning on graphs, a class of computation that does not perform well with traditional MapReduce frameworks [87, 105, 106]. GraphLab provides a programming model and an execution model for asynchronous, dynamic, graph-parallel computation. The programming model specifies a graph and functions on it. A data graph consists of $G = (V, E, D)$ consists of vertices, edges, and data. Update functions modify data associated with a vertex and schedule future updates on other vertices. Formally, an update is a function $f(v, S_v) \rightarrow (S_v, T)$ that consumes a vertex v and its associated scope S_v, which includes both its own data and data from its adjacent vertices, and produces an updated scope S_v and additional tasks T for future iterations in the analysis.

The execution model parallelizes updates on the graph by maintaining a list of tasks to perform on vertices. Workers compute in parallel, retrieving scheduled tasks from the list and placing new ones onto the list. GraphLab's run-time system can schedule tasks for performance by, for example, ordering tasks to maximize data locality and minimize network communication. Task (re-)ordering reveals parallelism and increases performance, but the extent that tasks can be re-ordered is dictated by the consistency model. GraphLab provides several consistency models, from full consistency, which requires concurrent updates on the graph to be at least two vertices apart, to vertex consistency, which allows any set of concurrent updates.

For example, consider the PageRank algorithm within the GraphLab framework. The algorithm recursively defines the rank $R(v)$ of a vertex v:

$$R(v) = \sum_{u|(u,v) \in E} R(u)j / \deg(u),$$

where $R(u)$ is the current rank of neighboring node u and $\deg(u)$ is the degree of that node. Recall that a graph $G = (V, E, D)$ consists of vertices, edges, and user-defined data. Suppose we compute PageRank for web pages, counting links between pages and identifying popular pages. The data graph is obtained directly from web links. Each vertex is a web page, each edge is a link, and each piece of data is the current estimate of a page's rank. GraphLab initializes the PageRank computation by enqueueing a task to compute pagerank for every vertex v in the graph. If the update function changes the rank of a vertex, the ranks for neighboring vertices must also change and the corresponding update tasks enqueue for future computation. Figure 2.8 illustrates the pseudo-code for the PageRank update function.

```
pageRank(v):
  R_old(v) = R(v)
  for each u in Neighbors(v)
     R(v) = R(v) + R(u)
  if |R(v) - R_old(v)| > eps
     queue pageRank(u) for each u in Neighbors(v)
```

Figure 2.8: Pagerank implementation within GraphLab, modified from [105].

GraphLab is a relatively mature starting point for researchers in computer architecture who wish to study graph analytics. But it is only one of several graph analysis frameworks. Satish et al. survey multiple frameworks [142] and find that GraphLab outperforms Giraph [5], a framework built directly atop Hadoop MapReduce. This survey also includes Combinatorial BLAS [26] and SociaLite [143], which are more specialized frameworks for sparse linear algebra and graph queries, respectively. Although these frameworks perform better than GraphLab, they may be less accessible to the lay user. Finally, the Spark framework for high-performance, iterative machine learning atop Hadoop provides extensions for graph analytics in the form of GraphX [6, 167]. Spark's ability to cache data, in memory, across loop iterations may improve performance and reduce communication between workers.

Computational Kernels and Data. Satish et al. identify four representative and recurring computational kernels for graphs—PageRank, Breadth First Search, Triangle Counting, and Collaborative Filtering—and we summarize them here to illustrate computation that is well suited for graph analytic frameworks.

- **Page Rank** identifies popular graph vertices by iteratively calculating the rank PR for each vertex i, which increases with the rank of its neighbors j, such that $PR^{t+1}(i) = \sum_{j|(j,i)\in E} PR^t/\deg(j)$.
- **Breadth First Search** is a classic graph traversal algorithm, which identifies the minimum distance between a starting vertex and other vertices in the graph. The traversal may be top-down, bottom-up, or a hybrid of the two [19]. After initializing distances to infinity, BFS calculates the distance for each vertex i based on the distance calculated for its neighbors j, such that $D(i) = \min_{j|(j,i)\in E} D(j) + 1$.
- **Triangle Counting** enumerates the number of triangles in the graph. A triangle consists of two neighboring vertices that share a third neighbor. Each vertex communicates its list of neighbors with all adjacent vertices. Each recipient calculates the intersection between its neighbors' lists and its own, such that $N = \sum_{i,j,k,i<j<k} E_{ij} \cap E_{jk} \cap E_{ik}$.
- **Collaborative Filtering** fits a model that estimates a user's rating for an item given ratings for other users and items. Suppose the matrix R quantifies the known ratings such that rows enumerate users and columns enumerate items. Collaborative filtering factors the matrix $R = PQ$ by solving the following minimization problem: $\min_{p,q} \sum_{(u,v)\in R}(R_{uv} - p_u^T q_v)^2$, where u and v are indices over users and items.

Programming models and run-time systems, such as GraphLab and GraphX, provide ample support for these and other kernels. These computational kernels are general and applicable to a variety of large networks. Leskovec et al. have collected large datasets, with graphs from social and web networks, as part of the Stanford Network Analysis Project [94, 95]. These graphs vary in size but are invariably sparse. For example, a Google+ social network has 107.6 K nodes and only 13.6 M edges.

2.6 ADDITIONAL CONSIDERATIONS

Multi-tier Workloads. In addition to the software frameworks presented in this chapter, conventional applications for enterprise computing continue to be relevant. The Transaction Processing Performance Council (TPC) provides relevant workloads for online transaction processing [155]. For example, its TPC-C benchmark simulates users as they issue queries and transactions against a database. Transactions are representative of those in electronic commerce and multi-tiered datacenters. In tier one, front-end clients run the TPC-C application and issue transactions. In tier two, back-end servers run the database management system. For experimental infrastructure, tier zero comprises machines that emulate human users and generate a mix of requests to exercise the system.

Similarly, the Standard Performance Evaluation Corporation (SPEC) provides workloads for enterprise computing, such as web-based order processing (SPECjEnterprise [151], SPECjbb [150]), file servers (SPEC SFS [149]). These workloads exercise different parts of a high-performance, commercial server and are especially relevant when benchmarking cache and memory hierarchy performance. SPECjbb is particularly interesting for its focus on multi-tier, Java business applications. In tier one, a transaction injector issues requests and services to the back-end. In tier two, a back-end server implements business logic. As in TPC benchmarks, tier zero comprises a controller that directs the execution of the workload.

Emerging Workloads. The provision of computing as a service expands the scope of datacenter workloads. Mobile devices increasingly rely on the cloud for personal assistance and contextual analysis. At present, these workloads compute on general-purpose server architectures. In the future, however, these workloads could benefit from hardware acceleration. Hauswald et al. benchmark analytical workloads for intelligent personal assistants (Sirius) and port them to varied platforms such as graphics processors and FPGA-based accelerators [66]. Similarly, as neural networks gain popularity for approximating sophisticated tasks, future datacenters may supply neural networks as a service and accelerate them [31, 65].

General-Purpose Workloads. Moreover, democratized access to cloud computing means that traditional, general-purpose benchmarks should play a role in datacenter research. Elastic cloud computing offers virtual machines to users who can launch arbitrarily diverse computation. Although this computation often deploys open source frameworks for distributed, task-parallel computing, it may also deploy more conventional applications such as those in the SPEC CPU [148], SPEC OMP, PARSEC [20], and SPLASH [163] benchmark suites.

Regions of Interest. This chapter presented software frameworks for datacenter computing—web search, map reduce, memory caching, and graph analysis. Computer architects should be wary of treating these frameworks as benchmarks. Rather, architects should identify computational kernels and regions of interest within the frameworks, whether they be specific queries in web search, a machine learning library atop map reduce, a specific operation on a key-value store, or a particular analysis on a social network. When benchmarking datacenter workloads, input data and the computation on that data are as important, if not more so, than the programming model and run-time system for that computation.

A framework's run-time system is often conflated with the software application of interest. For example, Hadoop and Spark support machine learning computation, but the framework itself also exercises the hardware, with computation to split input data, route intermediate results, and sort data at the reduce workers. A computer architect who benchmarks a Hadoop job might capture this computation for management in addition to the specific computation of interest. More precise measurements require hardware counter APIs or simulation checkpoints that are flexible enough to demarcate regions of interest in the application.

CHAPTER 3

Design

Domain-specific hardware design improves performance and power efficiency. Server architects favor modest hardware components for modern warehouse-scale datacenters, which prioritize total-cost-of-ownership and scaled out, rather than scaled up, performance. Architects have reduced cost and increased efficiency with their choices in processor microarchitectures and memory interfaces, which adapt design strategies from mobile and embedded devices for server platforms. Architects have focused on processors and memories because increasingly sophisticated mechanisms for memory caching shift activity to these resources. Yet, when memory caches are distributed across multiple machines, network latency also matters and architects have turned to accelerators for network interfaces. Finally, architects study rich design spaces to optimize individual components, but they must also organize these components into balanced and easily managed systems.

3.1 PROCESSORS AND COMPUTATION

Datacenter workloads exhibit two major forms of parallelism across tasks and across data partitions. For example, a search engine serves independent queries and calculates relevance scores for each server's share of the index with little communication. The natural parallelism in this system organization is well suited to warehouse-scale datacenters. In this setting, computer architects should differentiate between scale-out and scale-up processors, and they should explore the scale-out design space to optimize a combination of performance, power efficiency, and cost.

Commodity Designs. Barroso et al. make the case for web search on a cluster of commodity processors, which provides performance and reliability at modest cost [17]. Low-cost processors are particularly important when optimizing the total cost of ownership; a processor's capital cost is amortized over two to three years and should be kept low. Although commodity designs lack the expensive mechanisms for reliability and fast communication often found in more expensive server designs, these hardware mechanisms are rarely needed for datacenter software. Software-based approaches deployed atop commodity design can ensure reliability at scale. Moreover, datacenter workloads rarely employ fine-grained communication between processors and cores. Finally, Barroso et al. note that web search exhibits modest instruction-level parallelism, due to data-dependent control flow and dynamic data structure traversal. This observation means that out-of-order processors may be designed well beyond the point of diminishing marginal returns.

Small Designs. If out-of-order processors provide only modest performance, smaller or in-order processors might provide comparable performance at lower cost and greater efficiency. Smaller processors are particularly attractive for their power efficiency as processor power accounts for the largest share of server power [16, 138]. Smaller processor cores, typically characterized by shallower pipelines, narrower issue width, and in-order execution, dissipate much less power than their counterparts in traditional servers. However, small cores may improve efficiency at the expense of performance and service quality. Thus, any proposal to introduce small cores into a datacenter must precisely quantify these performance and power trade-offs.

In the early days of chip multiprocessor research, design trade-offs favored small cores for I/O-intensive datacenter workloads such as online transaction processing. Barroso et al. argue in favor of simpler cores (i.e., single-issue, in-order core) to reduce design effort and cost while delivering performance for database and web applications [18]. Davis et al. argue in favor of mediocre cores that use multi-threading to hide memory latency and increase throughput [33]. Kongetira et al. implement a highly multi-threaded chip multiprocessor, which trades instruction-level parallelism for thread-level parallelism, to produce the power efficiency needed by commercial server applications running in datacenters [80].

Lim et al. look beyond commodity design and argue in favor of low-end and mobile-class designs based on their total cost efficiency [101]. A server design incurs two types of costs—capital costs are incurred when the datacenter architect purchases the machine, and operating costs are incurred when she powers the machine. After combining these costs to estimate total cost of ownership (TCO), Lim et al. evaluate server designs according to performance per TCO dollar. Mobile and embedded hardware fare particularly well according to this metric.

Lotfi-Kamran et al. advocate a re-allocation of resources within a processor design, observing that area devoted to big last-level caches might be better used for additional cores [104]. Conventional designs have low performance density because aggressive out-of-order cores are designed past the point of diminishing marginal returns in performance and because large caches are less effective for datacenter workloads. Lotfi-Kamran et al. reduce last-level capacity per core and tightly integrate chip components into pods, which comprise cores, cache, and fast interconnect. As die area increases, architects could increase chip capability by scaling out the number of pods. This design strategy reduces core complexity and reclaims cache area for smaller cores, thereby improving performance density and power efficiency for applications such as MapReduce and web search.

The advent of ultra-low-power processors for phones and tablets (e.g., Intel's Atom and ARMs Cortex) fuels microarchitectural studies for web search in particular. Search engines are increasingly compute-intensive and a study of the Microsoft Bing search engine highlights the benefits and risks from using small processor cores [137, 138]. In this study, Janapa Reddi et al. compare and contrast Bing performance and efficiency on two very different machines, executing queries against a web index sized to fit in memory. One machine deploys conventional, four-core Intel Xeon processors, running at 2.5 GHz, issuing four instructions per cycle, and executing instructions out-of-order. The other machine deploys processors originally designed for mobile platforms. The dual-core Intel Atom processors run at 1.6 GHz, issue two instructions per cycle, and execute instructions in order. A clamp ammeter on the line that connects the voltage regulator and the processor measures power as Bing serves queries. As seen in Figure 3.1, an Atom core dissipates 1.5 W whereas a Xeon core dissipates 15 W. Naturally, such order of magnitude power savings tempt computer architects.

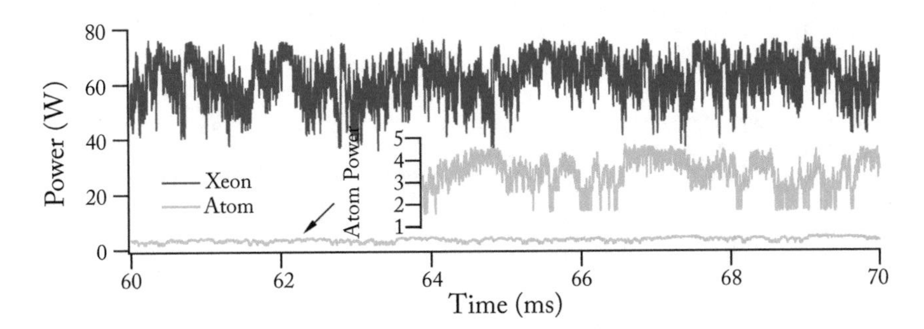

Figure 3.1: Physical measurements of processor power during Microsoft Bing search computation. These measurements of a four-core Intel Xeon and the two-core Intel Atom show that an in-order core dissipates 0.1× the power of an out-of-order core [137, 138].

Unfortunately, small cores harm query latency and service quality. Figure 3.2 illustrates the latency distribution for query streams on both machines. The mean and variance increases when Bing runs on the Atom. The horizontal axis reports latency that has been normalized to a cut-off time, which is the time allotted to a query before the search engine considers the query a failure. More than 99% of Bing queries complete before cut-off on the Xeon but only 90% of them do so on the Atom.

Moreover, Figure 3.3 shows that complex queries are disproportionately affected by performance limitations in the low-power processor. Suppose that the low-power Atom can sustain A queries per second and meet its quality-of-service target. As query load increases by multiples of x,[1] the percentage of queries that meet its performance target falls. This reduced service quality affects complex type C queries but has little effect on simple type A queries. Collectively, these measurements suggest that small cores can accommodate the majority of simple queries but big cores are needed to ensure performance for less common, complex queries.

The conclusions drawn from physical measurements of an industry-strength search engine are consistent with those from simulations of an open-source search engine [9, 11, 23]. Bragg et al. use Nutch to crawl and index Wikipedia pages, and use the Lucene search engine to serve queries. They use MARSSx86 and DRAMSim2 to simulate Lucene and estimate query response times for a broad processor design space [126, 140]. Designs in this space differ in clock frequency,

[1]Data from Microsoft Bing obscured for confidentiality.

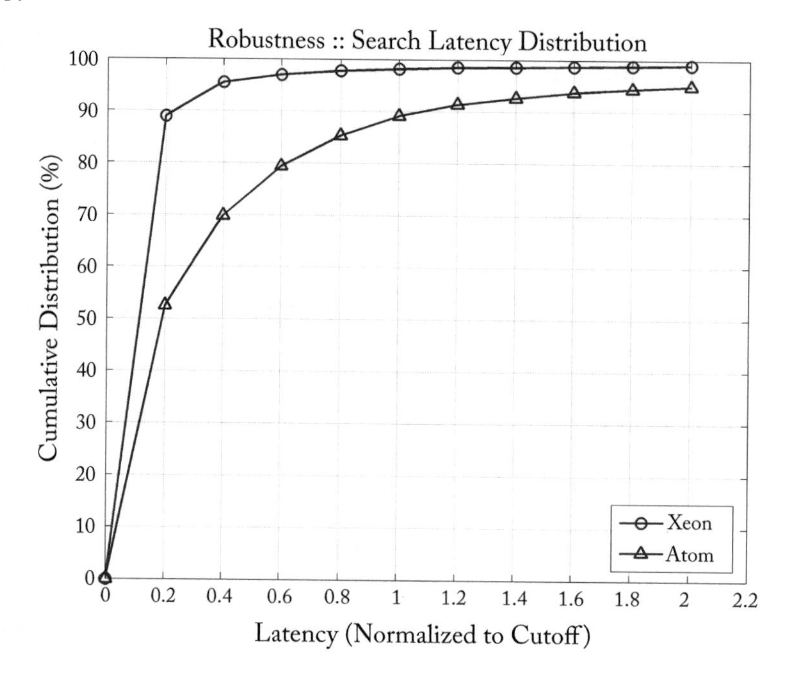

Figure 3.2: Query performance on Xeons and Atoms, illustrated with a cumulative distribution function on latency [137, 138].

superscalar width, and cache size. Table 3.1 summarizes the parameters, which define 18 in-order and 24 out-of-order core designs.

Table 3.1: MARSSx86 processor simulation parameters for a design space with 18 in-order and 24 out-of-order cores [23]

	In-Order	Out-of-Order
Frequency (GHz)	1 or 2	1 or 2
Superscalar Width	1, 2, or 4	2, 4, 6, or 8
L1 I-Cache	32KB, 4-way, WB	64KB, 4-way, WB
L1 D-Cache	32KB, 4-way, WB	64KB, 4-way, WB
L2 Cache	from 256KB to 1MB	from 1MB to 4MB
Power (W)	2.01 — 8.15	5.56 { 33.95

Figure 3.4 reports latencies for queries of varying types. Latency clearly depends on query complexity. Single-word queries complete quickly and latencies increase with query length, as seen for triple- and quad-word queries. The inverse single query is most expensive as it must search and

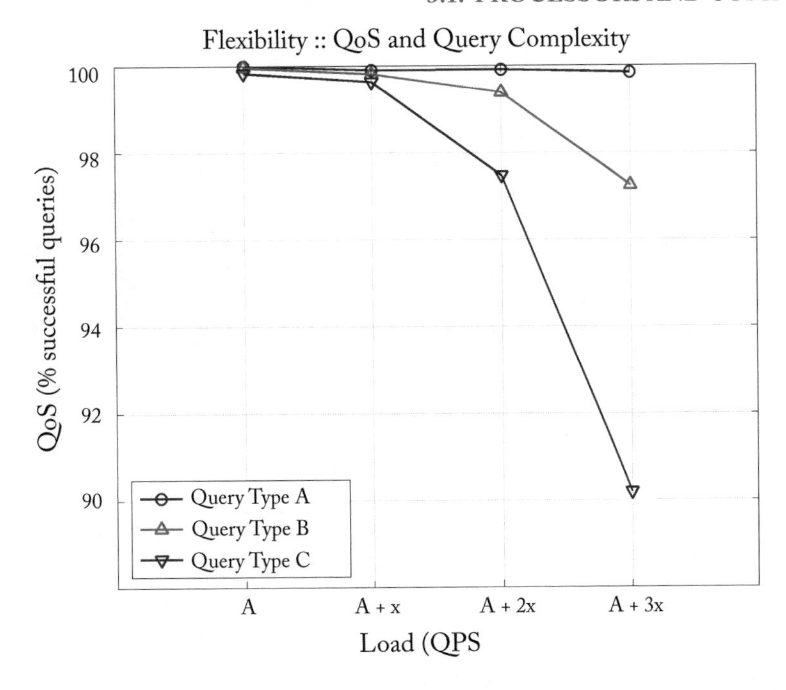

Figure 3.3: Query performance on low-power Atoms, illustrated with an analysis of service quality as query load increases for three query types. Type A corresponds to the simplest queries and Type C corresponds to the most complex [137, 138].

return the largest number of relevant pages. Average latencies for the 18 in-order cores and those for the 24 out-of-order cores show the impact of core design. Remarkably, the benefits of out-of-order execution are most prominent for the simplest queries; latency for single-word queries falls by more than 50%. In contrast, latency reductions for complex queries, such as inverse single, are more modest.

Better microarchitectures benefit complex queries less because the search engine breaks a complex query into multiple, simpler queries. For example, a search for "lucky rabbit" requires three steps—a search for "lucky," a search for "rabbit," and a step to integrate results. Larger, out-of-order cores handle load from multi-step query processing better. The benefits from out-of-order cores vary, but simulated performance for an open-source search engine is strikingly similar to measured performance for an industry-strength search engine.

In summary, despite the power efficiency, the case for small cores in datacenters is far from clear. Janapa Reddi et al. produce a qualified case for small cores—although small cores are sufficient for most simple queries, big cores may be needed for complex ones [137, 138]. Complex queries suffer from branch misprediction and limited cache capacity when computing in small cores. Microarchitectural limitations mean that, relative to big cores, small cores sustain 0.5× the

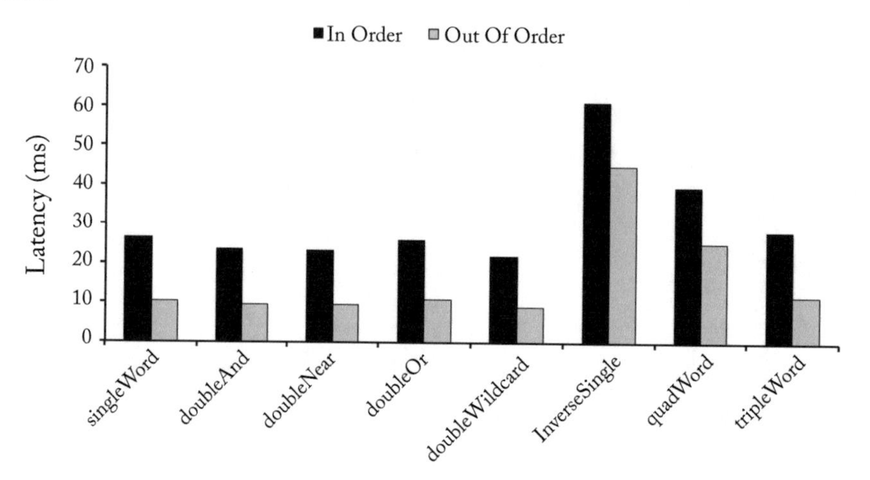

Figure 3.4: Simulated measurements report average response time. Measurements collected for in-order and out-of-order cores, and for a variety of query types [23].

query throughput and suffer 3× the latency. Yet, small cores dissipate 0.1× the power. The net result is a 5× improvement in energy efficiency, measured in queries per Joule.

Heterogeneous Designs. Hoelzle notes that Amdahl's Law may limit the adoption of small cores [68]. Although web search exhibits natural query-level parallelism, the limitations of small cores may require substantial intra-query parallelization or performance tuning to maintain performance. In other words, small cores may harm serial performance and may require compensating gains in parallel performance. The point at which small cores makes sense depends on the serial and parallel fractions of the application. Hoelzle argues against small cores once their serial performance trails big core performance by more than 2×.

The debate surrounding big and small cores motivates datacenters with a mix of core types. Heterogeneous processor design for datacenters mirrors prior work in heterogeneous core design for chip multiprocessors [83–85]. Multiprocessor architects focus on heterogeneous microarchitectures for a given instruction set architecture, a goal that is shared by datacenter architects. A common instruction set facilitates task migration across core types during software scheduling and hardware allocation.

However, a datacenter architect's view of heterogeneity differ from a multiprocessor architect's view with respect to integration and composition. First, whereas multiprocessor architects assume that multiple core types are integrated into a single chip, datacenter architects could construct a heterogeneous datacenter with several types of multiprocessors, each with homogeneous cores. Indeed, the degree of heterogeneous hardware integration remains an open question for datacenters—heterogeneous core types could be implemented within a socket, across sockets, across servers, or across racks.

For example, KnightShift is a heterogeneous server that integrates high-performance and low-power processors to improve energy proportionality [162]. The high-performance design's fixed power, which is dissipated regardless of utilization, is inefficient at low datacenter loads. In contrast, the low-power design is particularly efficient at low loads. To permit a run-time choice between the two designs, KnightShift provisions and networks both within a server.

Second, the heterogeneous composition of processor types is important, especially for warehouse-scale datacenters. Whereas multiprocessor architects study strategies for choosing core types for inclusion on chip, they rarely study strategies for choosing core ratios in a large system. In contrast, a datacenter deploys tens of thousands of processor cores and a decision to adopt heterogeneity requires a decision about the types of these cores. For example, if a datacenter were to deploy a mix of big and small cores, how many should be big and how many should be small? The answer would determine latency and throughput distributions as well as power efficiency.

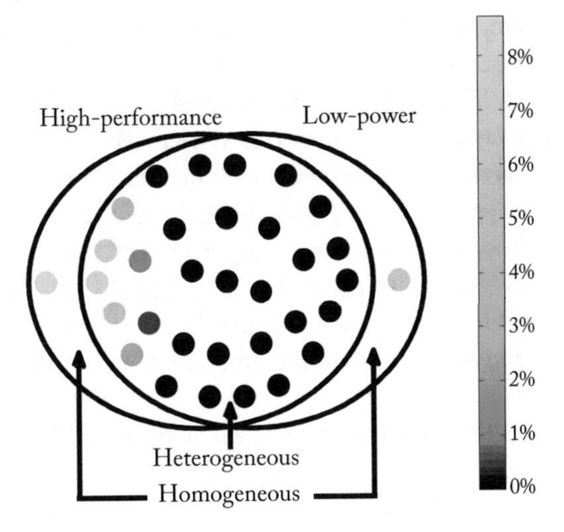

Figure 3.5: Venn diagram illustrates varied mixes of big and small core types given a datacenter's power budget. An ellipse represents a core type and a point in the Venn diagram represents a mix of types. Colors represent how often tasks violate service targets.

Guevara et al. propose datacenters that mix big and small cores. Moreover, they devise mechanisms to allocate heterogeneous cores to diverse tasks [58]. Specifically, Guevara et al. consider complex queries that prefer big cores and simple queries that are indifferent to core type. Given these preferences, the datacenter architect must optimize the datacenter's composition. Figure 3.5 illustrates the datacenter design space given high-performance and low-power cores. A datacenter that uses only high-performance cores violates service targets for 6% of its tasks because the architect cannot deploy a sufficient number of big cores to meet demand peaks given the facility's power budget. A datacenter that uses only low-power cores violates service targets because the mean and tail latencies are long, especially for complex tasks. In contrast, some heterogeneous mix of cores balances performance and power efficiency to ensure service quality.

Heterogeneous design introduces management risk, which should be anticipated during design space exploration. Guevara et al. provide a taxonomy for heterogeneous system design [59]. The taxonomy reveals design paths that balance performance, energy efficiency, and management complexity. Three types of risk may prevent a heterogeneous system from realizing its expected performance and power efficiency. First, application risk arises when architects use a benchmark suite during design and encounter a different set of applications during run-time. Second, system risk arises when architects expect one mix of applications during design and encounter a different mix of applications during run-time. Third, contention risk arises when applications compete for the same resource types, which causes some applications to compute on sub-optimal platforms. Among heterogeneous design strategies, those that are aware of management risk provide the best service quality when the resulting hardware is deployed in a cluster that serves diverse users.

In addition to heterogeneity from design, datacenter operators encounter heterogeneity from differences across hardware generations. Warehouse-scale datacenters deploy commodity hardware that is replaced periodically and inexpensively. Staged procurement leads to multiple microarchitectural configurations within the same system. Mars and Tang propose task mapping strategies that account for generational heterogeneity in datacenter workloads [109]. These strategies score software-to-hardware mappings, according to software preferences for microarchitectures and co-locations. Then the datacenter manager searches for mappings that produce better scores. Delimitrou and Kozyrakis also propose task mapping strategies that account for heterogeneity with collaborative filtering and a recommendation system [37]. Sparse profiles measure application performance on varied architectures and train a recommendation system that predicts performance and preferences.

Boosted Designs and Accelerators. Dynamic voltage and frequency scaling (DVFS) is a mechanism for throttling and boosting server performance. Computer architects have long studied DVFS for power savings and those strategies could apply to datacenter servers as well. For example, Lo et al. use DVFS to slow server computation, consume slack in the service level target, and reduce power dissipation [103]. Today's servers can enforce a fine-grained Running Average Power Limit (RAPL) and settings for this limit can be guided by profiles of latency slack in workloads such as web search and distributed memory caching.

In addition to throttling, DVFS can boost performance. Computational sprinting boosts performance by activating cores and boosting their voltage and frequency for a short period, which dissipates extra power and requires recovery to release accumulated heat [134, 135]. Applied judiciously, sprinting can reduce tail latencies in datacenter applications. Hsu et al. manages a processor's voltage and frequency according to memcached's set, get, and delete requests [70]. The mechanism relies on dual power supplies with two independent and external voltages for fast transitions between V/F levels. Higher V/F levels benefit some tasks more than others and a detailed characterization provides V/F transition rules to boost specific types of tasks and queries.

Beyond mechanisms for general-purpose processors, researchers have sought to improve web search efficiency through hardware specialization. Putnam et al. design and deploy a reconfigurable fabric, Catapult, for the Bing search engine to accelerate page scoring and double query throughput [131]. The Catapult fabric improves performance by synthesizing specialized logic that scores web pages onto a cluster of field-programmable gate arrays. The system improves performance yet avoids the design and fabrication costs of application-specific integrated circuits. Reconfigurability is particularly attractive as datacenter applications evolve rapidly—weekly updates are common [16]. Moreover, the system presents new and interesting management questions as varied software tasks request time and resources on the reconfigurable logic.

Reconfigurable logic is a waypoint on the path to custom accelerators, which promise even greater performance but at additional design cost [62, 63]. Kocberber et al. present a custom functional unit to support key-value stores [79]. Beyond the memory needed to hold data, distributed memory caching and key-value stores must perform hash table lookups. These lookups are amenable to acceleration and new hardware. The critical path in hash index lookups include ALU-intensive hashing and pointer chasing through a linked list. Kocberber et al. design on-chip functional units that accelerate this computation with pipeline and data parallelism. Pipeline parallelism increases throughput by separating an index operation into key hashing and linked list traversal. Data parallelism increases throughput by performing multiple index operations for multiple queries simultaneously.

Wu et al. design hardware accelerators for database workloads [164, 165]. For partitioning large datasets, a first step in more sophisticated queries, the authors design an accelerator for range partitioning that sits between the memory controller and processor core [164]. The hardware accelerated range partitioner (HARP) draws data from stream buffers in bursts, serializes the data, and performs a series of pipelined comparisons against splitter values that define range boundaries. For the queries themselves, Wu et al. propose database processing units (DPUs) comprised of heterogeneous, application-specific integrated circuits that quickly and efficiently process relational tables and columns [165]. The DPUs are programmed with coarse-grained instructions to increase parallelism, reduce register spilling, and amortize the overheads of instruction fetch. These acceleration strategies increase database performance by an order of magnitude or more.

Even with efficient hardware solutions, an even more efficient solution computes the "same" answer with less work. Baek et al. propose approximations for search that score fewer pages yet produce the same top results [15]. Controlled program approximation requires a calibration phase, which creates a quality-of-service baseline, and a dynamic phase, which determines whether an approximate version of a function produces a sufficiently good answer. For web search, software approximation reduces the time spent scoring pages, increasing query throughput and reducing the number of machines required to serve a given load. The difficulty in approximate computing, however, is the programming model—programmers must specify tolerable approximations and impose discipline on mechanisms that deviate from the precise baseline [43].

3.2 MEMORY AND DATA SUPPLY

Datacenter servers predominantly employ von Neumann architectures, which hold instructions and data in memory. Yet datacenter applications present unique challenges in cache and memory design. Instructions exert pressure on the cache hierarchy as the code's working set size is often larger than the L1 instruction cache and, sometimes, larger than the unified L2 cache [49]. As managing competition for cache capacity between instructions and data becomes increasingly important, architects can draw on a wealth of academic research and industrial design in cache partitioning; for examples, see set dueling research [132] and cache allocation technology in modern processors [71].

Data is similarly large and unwieldy, requiring substantial cache and memory capacity but relatively little bandwidth [82, 107]. Online and interactive services (e.g., web search) partition data to fit in a server's main memory and distribute partitions across multiple servers. Low-latency data backends use key-value stores to distribute memory caches. In this setting, computer architects seek lower average memory access time and higher energy efficiency.

Memory Bandwidth and Interfaces. Frameworks for distributed memory caching present several interesting observations that affect the way architects navigate the hardware design space. First, memory caching demands capacity but poorly utilizes bandwidth. Kozyrakis et al. find that datacenter applications use 65–97% of a server's DRAM capacity, but less than 10% of its peak DRAM bandwidth [82]. Applications partition data sets to fit in main memory. For example, Microsoft's Bing search engine partitions the web index into pieces and pre-fetches these pieces into memory to avoid expensive disk I/O. However, data popularity follows a Zipfian distribution and produce sparse, fine-grained accesses, which are unlikely to saturate memory bandwidth. Indeed, Kozyrakis et al. find that web search is constrained by memory latency, not bandwidth. Frameworks for memory caching are also unlikely to use much bandwidth. Network bandwidth is an order of magnitude lower than memory bandwidth (e.g., 10 GB/s from an Ethernet link vs. 12 GB/s from a DDRx bus) and presents a communication bottleneck.

The low demand for memory bandwidth motivates new memory interfaces. Delay-locked loops (DLLs) and on-die termination (ODT) ensure signal integrity when the data transfer rate is high, but these interfaces dissipate static power, which is inefficient when bandwidth demand is low. Figure 3.6 shows the power breakdown for a DDR3 chip that transfers data at 15% of its peak bandwidth. Only 14% of the total power is spent on satisfying requests from the memory controller—activating rows, placing data into a row buffer, and issuing read/write commands to the buffer. Termination power is dissipated by resistors that suppress signal reflections and ensure signal integrity but draw static current. Background power is dissipated by delay-locked loops and other interface circuitry that synchronize signals but dissipate power regardless of activity on the memory channel.

Malladi et al. find that LP-DDRx interfaces can improve energy proportionality by eliminating expensive circuitry from DDRx interfaces, which are used in today's datacenter servers [107]. Figure 3.7 illustrates the energy required per bit transferred as memory bandwidth utilization varies. DDRx interfaces are energy disproportional—the interfaces dissipate 70 pJ per bit at 100% bandwidth and 260 pJ per bit at 10% utilization. When large, fixed power costs in DDRx are amortized over a modest number of data transfers, energy per transfer increases and energy proportionality is compromised. As bandwidth utilization falls, static power dissipated by the interfaces is amortized over fewer bits and the cost per bit increases.

Note that the low voltage variant, LVDDRx, does not improve energy proportionality. Because static power dissipated by high-performance interfaces is the problem, simply scaling voltage and frequency can be detrimental. Ensuring signal integrity at low voltages and high data rates is more challenging than doing so at higher voltages. LVDDRx termination power per pin is higher than DDRx's. Moreover, lower voltages require corresponding reductions in clock frequency and peak bandwidth. In effect, adopting low-voltage interfaces incurs bandwidth costs without benefiting energy proportionality.

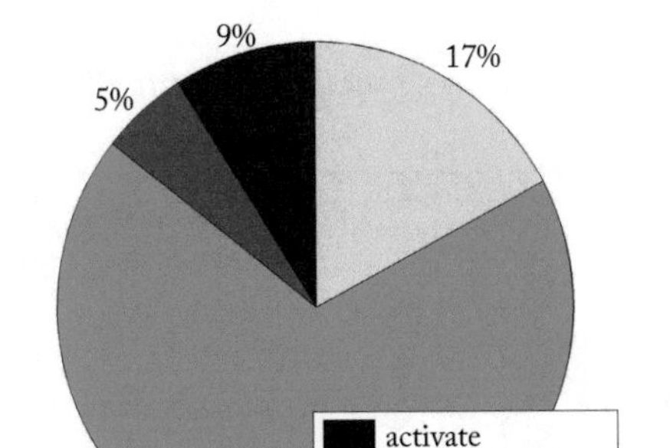

Figure 3.6: Power breakdown for DDRx channel under 15% utilization [107, 113].

The LP-DDRx memory interface, originally designed for mobile and embedded devices, forgoes expensive interface circuitry, thereby eliminating fixed power costs and improving energy proportionality. Figure 3.7 shows LP-DDRx's energy costs per transfer are comparable whether the memory channel operates at 10% or 90% of utilization. Without on-die termination and delay locked loops, LP-DDRx provides only half the bandwidth of DDRx. Yet Figure 3.8 shows that many datacenter applications, from web search to memcached, demand only a fraction of peak bandwidth. Compared to DDRx, LP-DDRx reduces memory power by 3–5× and improves energy proportionality without affecting performance.

Memory Bandwidth and Dynamic Scaling. In another strategy to tailor memory bandwidth to application demands, Deng et al. propose Memscale [39], which applies dynamic voltage and frequency scaling to the memory controller, and applies dynamic frequency scaling to the

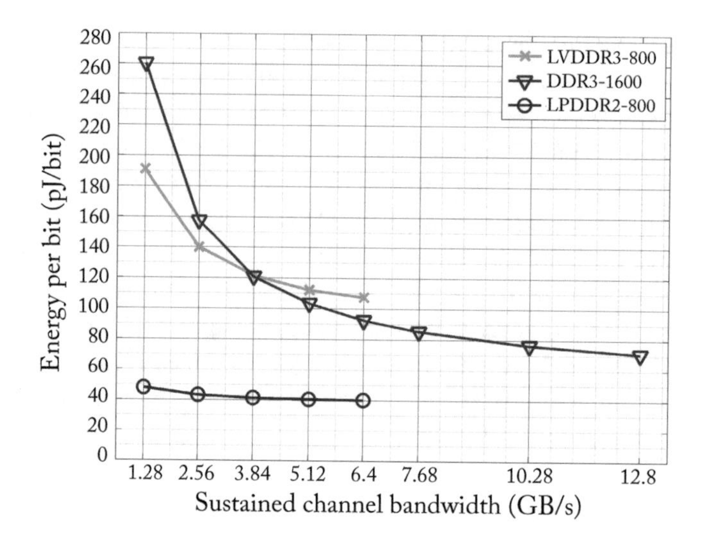

Figure 3.7: Energy per bit transferred, as a function of bandwidth utilization [107].

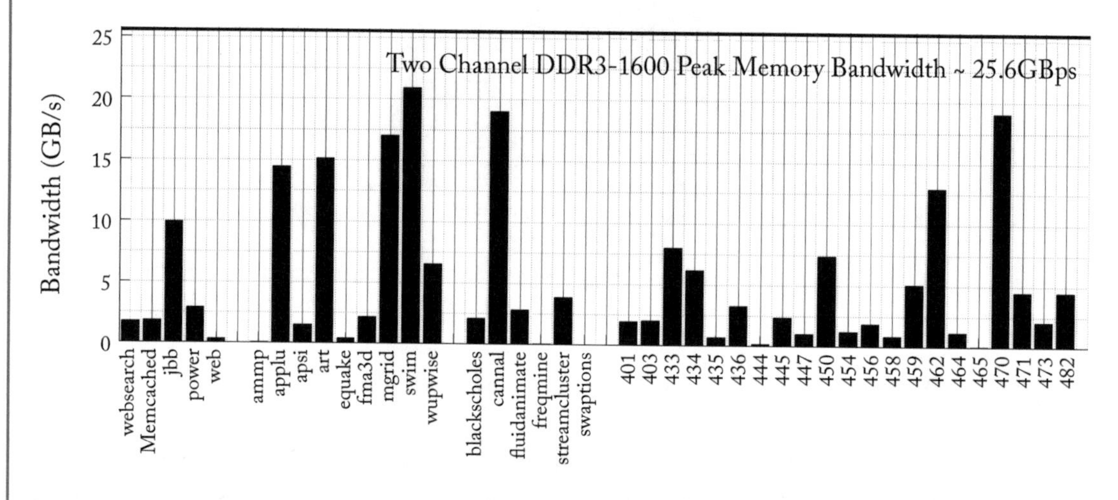

Figure 3.8: Bandwidth utilization for varied benchmarks, including datacenter workloads—web search, memcached, SPEC-jbb, SPEC-power, SPEC-web [107].

memory channels and DRAM devices. The Memscale control algorithm determines performance slack by using performance counters that track the amount of work pending at each memory bank and channel. It then adjusts the memory bus frequency by halting transfers and re-synchronizing the PLLs and DLLs. Memory bandwidth varies linearly with channel frequency, and lowering the frequency reduces power. Whereas prior studies identify idle DRAMs and enter deep sleep modes during which the DRAMS cannot serve requests [40, 46, 88, 124], Memscale is one of the first dynamic schemes for active, low-power modes, which saves power while serving memory requests to balance performance and efficiency.

Malladi et al. propose MemBlaze [108], which re-designs DRAMs and their links to enable fast power-up from deep sleep modes. MemBlaze removes timing circuitry from the DRAMs, circuitry which dictate the critical path during power-up, and shifts functional responsibility for timing to the memory controller. Specifically, the most efficient low-power modes turn off delay locked loops (DLLs) and clocks, but turning these circuits on during wake up requires a lengthy recalibration period. By eliminating these circuits from DRAM chips, and by adding clock and data recovery circuitry to the memory controller, Malladi et al. present a memory architecture in which DRAMs can sleep and wake quickly. MemBlaze permits data transfers immediately after power-up, which produces many more opportunities to exploit deep sleep modes. MemBlaze reduces energy per transfer by 50% for memcached, SPEC JBB, and other benchmark suites for cloud applications.

Alternatively, Malladi et al. propose DRAMs that retain their timing circuitry (DLLs) and turn them on/off aggressively with MemCorrect and MemDrowsy mechanisms [108]. Aggressively power gating the timing circuitry may introduce synchronization errors. With MemCorrect, the memory controller speculatively transfers data immediately after wake-up, detects errors that arise when DLLs are not yet synchronized, and re-transfers correct data after synchronization errors. With MemDrowsy, the memory controller speculatively transfers data immediately after wake-up but at a slower rate. As DLLs synchronize and timing becomes more precise, the controller increases transfer rate. MemCorrect and MemDrowsy reduce energy per transfer by 50% for memcached, SPEC JBB, and other server workloads with only a 10% performance penalty, which is incurred when correcting timing errors.

Memory Capacity. Although datacenter applications demand little bandwidth, they do demand capacity. Architecting big memory servers is complicated because we must ensure signal integrity on the memory channel, which is a shared, parallel bus. We increase capacity by adding DRAM chips to the memory channel. But every additional chip and its contacts to the channel adds an impedance discontinuity, which generates reflections as signals travel between the memory controller and the DRAMs. In DDRx interfaces, on-die termination suppresses signal reflections, delay-locked loops synchronize signals to ensure timely samples from command and data buses, and the system architect limits the number of DRAMs that share the channel.

In eliminating synchronization and termination circuitry, LP-DDRx is more vulnerable to noise. Yet, Malladi et al. show how stacked dies within a package can reduce the number of pins that load the memory channel and permit four memory ranks per channel, which produces capacities that are competitive with those in today's servers [107]. To provision additional capacity, architects employ buffers that re-time signals to ensure integrity. The buffers are positioned between the channel and the DRAM chips. Buffers and DRAMs communicate via serial links to ensure reliable communication even as capacity increases.

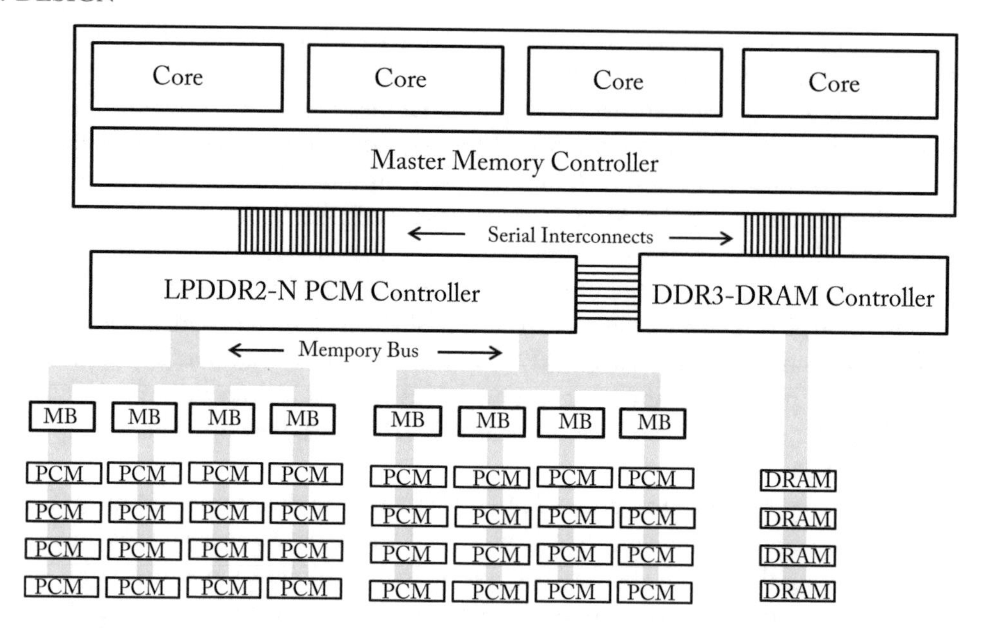

Figure 3.9: Hierarchical memory buffer organization that connects the processor to buffers with serial links and connects the buffers to DRAMs with parallel buses. [61].

Ham et al. take these ideas further by proposing a hierarchical organization for memory controllers and buffers to support heterogeneous and high-capacity memory systems [61]. Today's architects integrate the memory controller onto the processor die, which constrains the architecture of heterogeneous and high-capacity memory systems. First, the integrated controller may have difficulty accommodating diverse protocols and schedulers for different technologies. Second, the controller interfaces to parallel memory buses, which provide less bandwidth per pin than serial point-to-point links. Together, these limitations motivate disaggregated memory controllers with on-chip masters and off-chip slaves. Instead of implementing protocol commands, the on-chip master controller packetizes memory requests and delegates protocol implementation to off-chip slave controllers. The master communicates to the slaves via fast, serial links. The slaves send commands to buffers via parallel buses.

Figure 3.9 illustrates a hierarchical organization for memory buffers, which permits tens of ranks and hundreds, if not thousands, of gigabytes. Moreover, it shows how disaggregated controllers support heterogeneous memory technologies and protocols. The master memory controller processes last-level cache misses and forwards requests to the appropriate slave controller according to the data placement policy. For example, DRAM might be used as a cache for PCM and all requests query DRAM first. The slave memory controllers implement technology-specific protocols (e.g., DDRx for DRAM and LPDDRx-N) and access memory ranks via memory buffers and parallel buses. The buses are carefully loaded to ensure signal integrity—this example assumes that each bus accommodates four elements (e.g., memory buffers or ranks).

Blade servers present a different architectural strategy for memory capacity. Servers with distributed memory request data from remote memory with remote procedure calls (RPC) or remote direct memory accesses (RDMA). But recent research proposes blades that share an address space and communicate via a backplane interconnect such as PCIe.

Bridges and switches perform address translation and route memory requests to the appropriate blade. A processor can access a remote blade's memory with load/store instructions via PCIe. The memory controller, PCIe root complex, and non-transparent bridges are all integrated into the processor die. The memory controller identifies remote memory requests, diverts those requests to the root complex, and relies on bridges for address translation into the destination blade's physical address.

Lim et al. make the case for architecting memory blades specifically to supply memory capacity for compute blades [99, 102]. Hou et al. prototype a blade server that shares distributed memory via PCIe bridges [69]. Novakovic et al. present an architecture for low-latency, in-memory processing atop distributed shared memory [116]. Wang and Lee derive parameterized performance and power models to analyze non-uniform memory access in blade servers [160].

Memory Persistence. In the future, emerging memory technologies may further increase system capacity. Phase change memory, and other resistive memories, promise scalability with smaller devices that reduce programming currents [91, 133, 177]. Whereas a DRAM element is comprised of a storage capacitor and an access transistor, both of which offer difficult scaling trajectories, a phase change memory element is comprised of a chalcogenide that can be programmed with progressively smaller electrical currents as the device shrinks [92, 93]. PCM read performance is competitive with that of DRAM; PCM write performance is poor but datacenter applications (e.g., memcached) write much less often than they read. Moreover, PCM non-volatility offers fast persistence that may be valuable for storage systems.

Key-value stores, like RAMCloud [122] and FAWN [2], may find that resistive memories aid the pursuit of durability, availability, and consistency—first-class functional objectives. Byte-addressable, persistent memory permits fine-grained writes to storage, a capability that is not possible from disks that are accessed via an I/O controller. To exploit emerging memories, Condit et al. re-think file system design with BPFS, a filesystem that exploits atomic, fine-grained updates to persistent memory to improve reliability and performance compared to traditional file systems [32]. BPFS permits re-ordered writes within epochs, demarcated by barriers, but ensures write order across epochs. This notion of persistency balances write order and performance. Pelley et al. further explore these trade-offs with a framework to reason about stronger and weaker persistency models [129].

Memory persistency would address challenges in key-value stores that are currently addressed in system software with slower technologies. RAMCloud ensures durability with buffered logging [122], which stores each data object in a single memory location and logs updates to the object in multiple servers. RAMCloud logs updates in memory and transfers them to disk asynchronously, which increases disk bandwidth. Similarly, FAWN nodes buffer writes and updates until they receive acknowledgments from the log-structured datastore [2]. Resistive memories, such as phase change memory, reduce the need for such mechanisms as updates become durable far more quickly.

Although this lecture describes the potential of fast, non-volatile memory from the perspective of phase change memory, datacenter architects should be wary of placing bets on any one technology. Emerging memory technologies face several challenges, including their durability under repeated writes, their power and energy efficiency, and their fundamental scalability to smaller devices and larger arrays. Phase change memory, spin-torque transfer magnetoresistive RAM, and memristors are competing technologies. Yet at least one of these technologies will likely provide qualitatively new building blocks for big memory servers given the depth and breadth of ongoing research in non-volatile memories.

Balanced Design. Hardware design for datacenter workloads must balance the system, matching hardware capability to software intensity. Hardware capability is defined by the processor's performance, measured in operations per second, and the memory system's performance, measured in bytes transferred per second; the ratio defines the hardware's balance measured in operations per byte. Similarly, software intensity is defined by an algorithm's compute and memory intensity, measured in the number of operations performed for every byte transferred from memory. In a balanced system, hardware capability matches software intensity [86]. A balanced architecture must first ensure that bytes are transferred from main memory and not disk.

Distributed memory caching illustrates the importance of balanced design. Memcached servers that deploy high-performance, out-of-order processors risk overprovisioning computation relative to memory capacity. Gutierrez et al. study unconventional server processors and emerging memory technologies to right-size computation to capacity ratio [60]. Using low-power ARM cores correct for over-provisioned computation, replacing conventional DRAM with NAND Flash increases cache density, and 3D-stacking DRAMs integrate processors and network interfaces with memory. Li et al. study modern platforms and demonstrate high request throughput when processors, memory, and network subsystems are balanced [97]. Specific hardware mechanisms—for example, direct cache access, multi-queue network interface cards, prefetching—play key roles in sustaining one billion key-value requests per second in a balanced system.

Many other datacenter workloads already take great care and use memory judiciously for indexed data in web search or cached data in Spark. Surprisingly, graph analytics are a prominent class of datacenter workloads that do not require much memory capacity. The Stanford Network Analysis Project provides tens of graphs from a variety of domains, including social networks, online communities, and web graphs [94]. Most graphs consist of fewer than a million, sparsely connected vertices. Similarly, other real world datasets, such as those from Facebook, Wikipedia, Netflix, are often small enough to fit in a single machine's memory system [142]. Two datasets are larger—a Twitter graph consists of 1.5 billion edges and a Yahoo Music graph consists of 1.0 billion users.

However, even large graphs do not require much memory capacity—the Twitter graph is just over 30 GB in size, which is easily accommodated. Today's commodity servers could supply 512 GB in DRAM capacity with, for example, 8 GB per memory rank, four ranks per channel, four channels per processor, and four processors per server. If more memory capacity were required, servers could employ a hierarchy of buffers [61], which re-time signals to ensure their integrity and increase the number of memory channels. Thus, architects may pursue multiple strategies for memory capacity, which ensures the graph fits in memory. If the graph is too large and memory capacity is insufficient, data is distributed across multiple nodes and network bandwidth is the performance bottleneck.

Supposing the graph fits in memory, a system architect should consider memory bandwidth—the number of bytes transferred per second into the processor. Memory bandwidth is often the hardware constraint in graph analysis. Computational kernels stream through graph vertices and perform a modest amount of computation to each. For example, for every vertex loaded from memory, PageRank performs an add and a divide, and breadth first search performs an add and a compare. For such low ratios of operations to bytes transferred, modern servers may over-provision processor operations and under-provision memory transfers for graph analysis. Computer architects can either balance the system or introduce processors-in-memory, which might reduce memory bandwidth demand by shifting computation (e.g., summation) to the DRAMs. Fundamentally, however, software architects are best positioned to accelerate graph analysis with new data structures and algorithms that reduce the communication-to-computation ratio and produce the same answer with less work.

3.3 NETWORKING AND COMMUNICATION

The network presents a communication challenge within datacenters. Network performance dictates the critical path in key-value stores, which supply data from memory via the datacenter network. It is also prominent in graph analytics, which move data and tasks with varied granularities that depend on the application's chosen consistency model. For these and other applications, datacenter research must be cognizant of network bandwidth and latency.

Network bandwidth is limited and is at least an order of magnitude less than memory bandwidth [107]. For example, 10 Gb Ethernet supplies 1.25 GB/s of bandwidth whereas a DDR4 memory channel supplies 25.6 GB/s. Constrained bandwidth has implications for distributed memory caching in which get requests communicate data across memory channels and network links. A system is likely to saturate network links before saturating processor and memory resources within the backend server.

Network latency is a second challenge. Whereas processors and memories operate on timescales of nanoseconds, networks operate on timescales of hundreds of microseconds. The network is increasingly the dominant contributor to round-trip latency. Memory can supply data in one hundred nanoseconds but the network stack adds another one hundred microseconds. The principle sources of latency include the network stack within the operating system, which is designed for generality and not latency, the network interface card, and the network switches that connect racks in large datacenters [116, 141]. Latencies in each of these components extend the critical path, especially when distributed memory caches supply data at fine granularities (e.g., KBs) and network overheads are amortized over small payloads.

The network interface dictates the overheads from distributed memory caching. Software implementations of the TCP/IP stack for Ethernet, a commodity technology, are expensive, and researchers have considered a number of strategies. First, Nishtala et al. turn to UDP, a simpler protocol without support for reliable transmission, for lower transmission delay in Facebook's memcached deployment [115]. Moreover, Facebook employs flow control, restricting the number of outstanding client requests to limit incast congestion.

Second, Ousterhout et al. note that specialized networks, such as Infiniband and Myrinet, can support remote procedure calls with round-trip latencies of less than 10 microseconds [122]. Specialized network switches can further reduce end-to-end transmission time in large datacenter networks. In contrast, commodity networks, such as Ethernet, support calls with latencies between 300 and 500 microseconds. Unfortunately for memcached, datacenters are more likely to deploy commodity networks, despite the speed of specialized ones, to optimize total cost of ownership.

Third, network communication could bypass the operating system and reduce system software overheads. Much of the latency in networking is due to overheads in the operating system's network stack, which trades performance in favor of generality. Bypassing the stack reverses the trade-off. Ousterhout et al. proposes mapping network interface cards (NIC) into the application address space, which would eliminate context switches and shift responsibility for polling the NIC to the application [122]. Lim et al. propose dual accelerators for networking (i.e., TCP/IP offload) and hash index lookups to improve performance and efficiency [100]. These strategies can reduce round-trip latency by an order of magnitude or more.

CHAPTER 4

Management

Datacenters are large and complex systems that present unique management challenges. At scale, datacenter systems deploy many capable servers with multiple cores and abundant memory capacity. First, the datacenter allocates hardware in response to diverse requests from users and their applications. Second, the datacenter co-locates multiple software tasks on a single server to increase utilization and efficiency. Although their energy proportionality have improved, server power is still comprised of a large fixed component (e.g., 30–50% of peak power) and a variable component that increases linearly with utilization. An operator amortizes a server's fixed power over more work by co-locating multiple tasks within a server.

Datacenter management frameworks allocate hardware and co-locate software to balance multiple competing objectives. Performance is measured from several perspectives, all of which matter in some context. Throughput measures system capacity and affects the ability to respond to diurnal and peak activity patterns. Latency measures response time and dictates user experience. The latency distribution measures response time percentiles to ensure that the dominant majority of tasks (e.g., 99% of tasks) complete before a domain-specific target. The latency distribution exhibits a heavy tail—a few tasks report latencies much greater than the median. Finally, service-level agreements and obligations combine throughput and latency to produce a composite measure of welfare, expressed in a currency such as virtual tokens or physical dollars.

4.1 MANAGEMENT FRAMEWORKS

Datacenter operators manage resources at multiple levels and define clean abstractions between those levels. With abstraction, management frameworks can provide mechanisms that enforce a broad spectrum of policies. For example, the framework can provide mechanisms for the users to request resources and mechanisms for the system to grant resources, both of which would help implement policies for diverse objectives such as performance and fairness. By cleanly separating mechanisms from policies, datacenter operators permit independent innovation in both.

Moreover, abstraction separates multiple levels of management. Datacenter-level management allocates system resources to disparate software frameworks and run-times. For example, a datacenter that runs Hadoop, Spark, and GraphLab jobs simultaneously must allocate processors and memory to these jobs. Framework-level management allocates a job's resources to its constituent tasks. Each framework can manage its task queue differently. For example, Spark constructs a data dependence graph between tasks in its job and enforces these dependences when assigning tasks to hardware. In contrast, GraphLab examines dependences between graph vertices and relaxes these dependences according to the job's consistency model when assigning tasks.

Mesos is a management platform that accommodates multiple software frameworks with a two-level mechanism [67]. The Mesos master coordinates Mesos slaves, each of which runs on a datacenter node. The master distributes resources, such as processors and memory, to slaves. Each slave employs a framework-specific run-time system (e.g., Hadoop, Spark, MPI) to schedule tasks on resources received from the master. For example, Figure 4.1 illustrates the offer mechanism. When a slave completes tasks and jobs, it frees and returns resources to the master (e.g., 4 processors and 4 GB of memory). The master recovers these resources and, according to datacenter policies that further management objectives such as fairness, offers them to another slave. If another slave accepts the offer, its run-time system will schedule tasks on the allocated resources (e.g., 2 processors and 1 GB of memory to task 1, 1 processor and 2 GB of memory to task 2).

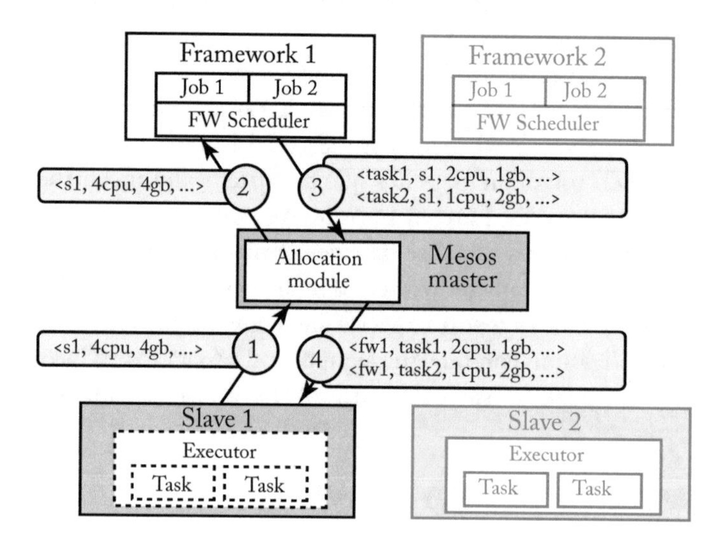

Figure 4.1: The Mesos master allocates datacenter resources to slaves, which deploy users' applications and run-time systems. Each slave allocates its resources to tasks within the application. Reproduced from Hindman et al. [67].

The Mesos master enforces datacenter-level policy and Mesos slaves enforce diverse framework-level policies. This separation of roles increases scalability as fine-grained task management is supplemented by less expensive, coarse-grained datacenter management. Moreover, this separation accommodates diversity as users deploy increasingly sophisticated and customized run-times for their applications. In return for these benefits, the datacenter may become less efficient as decentralized scheduling likely performs worse than globally coordinated scheduling. Note that the Mesos master does not account for users' resource requirements. Instead, the master offers available resources, and it asks a user to accept or reject the offer.

4.2 PROFILING AND CHARACTERIZATION

Performance profiling and analysis are prerequisites for effective datacenter management. At scale, datacenter profiles provide a rich dataset for characterization and inference. Coordinating, managing, and querying a repository of profiles at scale is difficult. Ren et al. describe Google's approach to datacenter-wide profiling [139]. Google-wide profiling (GWP) samples machines and jobs for profiling, ensures that profiling is non-intrusive, and aggregates profiles into an easily accessible database. The datacenter deploys continuous system profilers, such as OProfile [119], and hardware performance counters on each node. GWP is a distributed service that samples nodes in the fleet and events in the profiler to manage performance overheads. As GWP profiles the system, it links each profile to identifiers in a machine database and a binary repository to precisely identify the hardware-software conditions that produced the profile. Each profile is written to Google's distributed file system.

Google-wide profiling supports workload characterization at scale. For example, Kanev et al. use GWP to understand application trends and identify opportunities for further study [77]. Google's datacenter workloads are increasingly diverse and no single application presents a large optimization opportunity. The top 50 hottest codes consume only 60% of the datacenter's processor cycles. On the other hand, Google's shared libraries may present an opportunity as 30% of the datacenter's processor cycles are spent on six utilities. Whereas datacenter benchmarking can deploy a few workloads for detailed microarchitectural characterization, GWP draws on millions of tasks over months of continuous operation. Thus, datacenter-scale studies provide a new perspective and highlight the challenges of finding a single strategy to appreciably improve system performance at scale.

4.3 PERFORMANCE ANALYSIS

As profilers supply a wealth of data, computer architects require increasingly sophisticated strategies to translate that data into decisions. In this section, we describe three examples of such strategies—regression models, recommendation systems, and sensitivity analysis.

Regression Models. Statistical inference and machine learning supports big data analysis at scale. As system profilers sample tasks and events, they should sample to produce diverse and broadly relevant datasets. Wu and Lee find that sharded and portable profiles facilitate statistical analysis [166]. First, profilers should divide a software task into short shards so that new tasks can be understood in terms of multiple tasks' shards. With sharding, an application's profile will be useful in more scenarios. Second, profilers should measure microarchitecture-independent events so that a task's profile on one machine is relevant on another machine. For example, the profiler should measure re-use distances rather than cache miss rates.

With sharded and portable profiles, Wu and Lee statistically infer performance models that predict interactions between diverse hardware-software pairs [166]. They use regression models that predict performance $z = F(x, y) + \epsilon$ as a function of hardware parameters $x = x_1, \ldots, x_m$

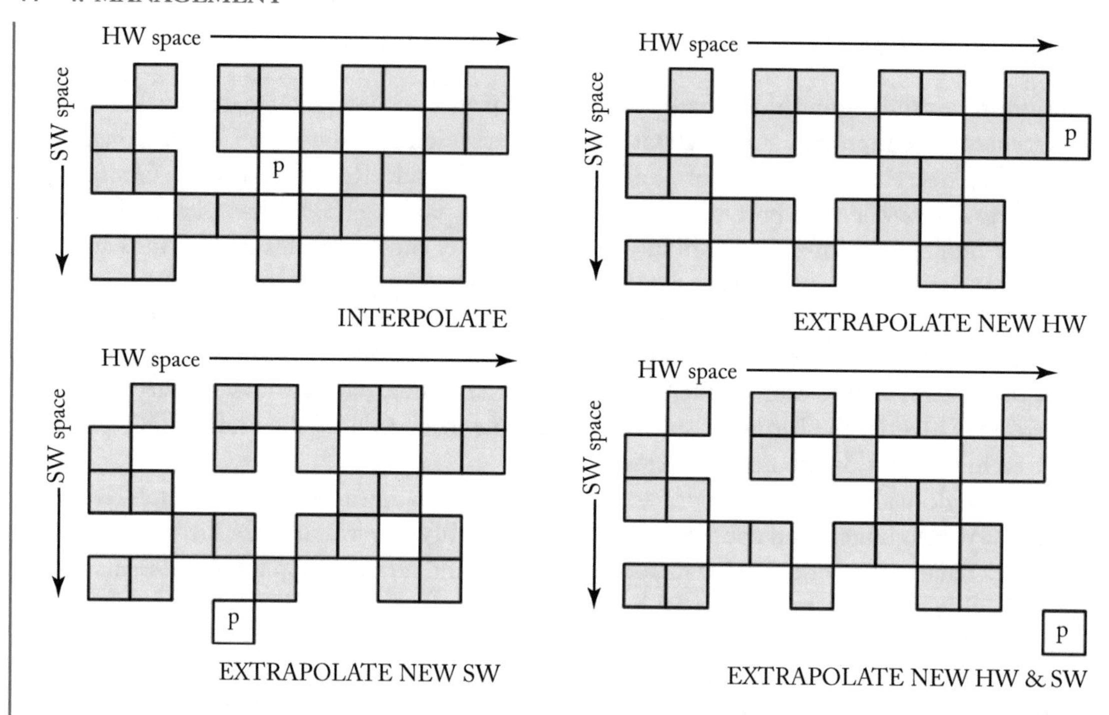

Figure 4.2: Sharded and portable profiles support statistically inferred models. With sufficient diversity in the training data, models can interpolate and extrapolate [166].

and software behaviors $y = y_1, \ldots, y_n$. Splines capture non-linear effects, transforming a model input x into $S(x) = \beta_0 + \beta_1 x_1 + \beta_2 x^2 + \beta_3 x^3 + \beta_4 (x - a)_+^3 + \beta_5 (x - b)_+^3 + \beta_6 (x - c)_+^3$ to increase flexibility at points a, b, and c in the range of values for x. Given hardware-software interactions between x and y, regression predicts performance with a product term in the model such that $z = \beta_0 + \beta_1 x + \beta_2 y + \beta_3 xy$.

As shown in Figure 4.2, by applying these regression strategies to sparsely sampled software and hardware profiles, we can interpolate to predict performance for previously unobserved hardware-software pairs, or extrapolate for new hardware or software. Extrapolation is particularly helpful when predicting the performance impact of a new processor microarchitecture, back-end compiler optimization, or data input for an existing software task.

Although regression is not typically effective for extrapolation, sharded and portable profiles facilitate prediction beyond the space defined by the training data. A new software application may not resemble a previously observed one, but its shards may resemble a combination of previously observed ones. Similarly, a new hardware architecture may resemble a mix of previously observed platforms, which supply portable training data. Thus, looking beyond monolithic applications and specific architectures permits more accurate prediction.

Collaborative Filtering. Recommendation systems use sparsely measured user-item ratings to recommend items to users. Whereas content-based filtering relies on specific user and item attributes (e.g., author, genre), collaborative filtering relies on implicit interactions between items, between users, and observed ratings. With a sufficiently diverse data set, collaborative filtering can produce a recommendation system that accurately suggests new and effective user-item pairs [81].

Profilers produce a sparse matrix with ratings for sampled user-item pairs. Collaborative filtering estimates the corresponding dense matrix with ratings for every user-item pair. Specifically, the sparse matrix with sampled and profiled ratings is factored such that user u's rating for item i is expressed as $r_{u,i} = q_i^T p_u$, where q and p are learned using stochastic gradient descent or alternating least squares. Given estimates for q and p, we have estimates for ratings $r_{u,i}$ for every user-item pair.

Delimitrou and Kozyrakis apply recommendation systems to datacenter management [37, 38], taking applications as users and execution environments as items. Each environment is defined by hardware parameters and co-located users. Their management framework separates these two effects on users by constructing one recommendation system for each. The first recommender analyzes a sparse matrix with items that correspond to heterogeneous machine configurations. The second analyzes a sparse matrix with items that correspond to varying degrees of interference from co-located users. The two recommenders estimate user preferences for hardware platforms and software co-runners, which can drive allocation heuristics.

Sensitivity Analysis. Datacenters co-locate workloads to improve energy efficiency by amortizing fixed power costs over more computation. However, co-location can harm performance as workloads compete for shared resources and interfere with each other's computation. Reasoning about co-location and performance is difficult—the basic approach requires w^n profiles where n is the number of workloads that share a machine and w is the number of unique workloads in the system. Yet targeted profiling can assess workload sensitivity to co-location without exhaustive profiling.

Mars et al. propose the Bubble-Up framework to assess each workload's sensitivity and contribution to resource contention in a shared machine [110, 168]. Suppose two workloads can share a machine. Rather than measure many workloads' pairwise performance characteristics, which is intractable, Bubble-Up performs a series of simpler measurements. First, the framework assesses a workload's sensitivity to contention by co-locating it with a bubble, which expands and contracts to reveal performance penalties under varied scenarios. Second, the framework assesses a workload's contributions to contention by measuring its utilization of shared resources such as last-level cache and off-chip memory bandwidth. Finally, Bubble-up estimates a workload's performance by mapping its co-located workload to a representative bubble and evaluating its sensitivity profile given that bubble.

4.4 RESOURCE ALLOCATION FOR PERFORMANCE

Performance analysis guides resource allocation. Datacenters supply a wealth of resources to be divided amongst users. At present, each user receives an allocation of processor cores and main memory capacity. In the future, they may receive fine-grained allocations of last-level cache capacity and memory bandwidth. The precise size of these hardware allocations depends on the management objective (e.g., throughput, latency, or fairness).

Resource allocation is trivial in some datacenters. Infrastructure-as-a-service provides users a menu of hardware platforms, each provisioned with specific resources and priced accordingly. From the datacenter operator's perspective, the management objective is revenue and the management mechanism simply responds to user requests. In this setting, the question of resource allocation is placed squarely on the user's shoulders. She must determine the number of resources required to meet her performance objectives. And she receives whatever resources she requests as long as she can pay the listed price.

Existing cloud computing frameworks place a prohibitively heavy burden on users, the majority of whom have little experience in performance profiling and computer systems. Users are likely conservative and request more resources than they require to meet performance goals, increasing datacenter inefficiency. Alternatively, they lack sufficient expertise and request fewer resources than they require. To address these challenges, researchers have devised new frameworks that couple software performance analysis with hardware resource allocation to achieve varied objectives.

Meeting Performance Targets. First, consider allocation mechanisms that build on the analysis frameworks presented thus far. Using a recommendation system to assess user preferences for hardware platforms and co-location scenarios, Paragon and Quasar allocate resources with heuristics that optimize user utility [37, 38]. Maximizing utility means minimizing interference and contention from co-located workloads and then maximizing performance from available server configurations. Paragon and Quasar are significant advances over existing interfaces for infrastructure-as-a-service as it hides the complexity, heterogeneity, and co-location effects within the datacenter. Instead of asking users to request hardware, Paragon infers their hardware preferences and allocates hardware accordingly to maximize throughput subject to latency constraints.

Similarly, Bubble-up uses a greedy allocation mechanism built atop its contention analysis framework [110, 168]. The datacenter operator defines a threshold for each application's tolerable performance degradation. Bubble-up uses its models, which capture each application's sensitivity and contributions to contention, to predict whether a given co-location will violate the threshold. The datacenter permits co-locations that preserve service quality.

Service-Level Agreements and Economic Mechanisms. Beyond direct measures of performance, datacenter operators can employ service-level agreements (SLAs), which specify user value that is derived from computation. The canonical SLA is a piecewise-linear function in which a user derives some constant value if latency targets are met and derives linearly decreasing value as latency exceeds the target. Conversely, value increases with task performance until the user experiences diminishing marginal returns from further performance. Value is often measured in a physical currency (e.g., dollars) and latency is often measured in percentiles (e.g., 95th percentile). SLAs provide a sophisticated interface for datacenter hardware. Instead of asking users to request hardware, SLAs ask users to specify their value for performance.

Because SLAs express value for performance, markets provide natural mechanisms for optimizing hardware allocation. Economic mechanisms for managing shared hardware date back to 1968 and Sutherland's market for time slots in a shared PDP-1 at Harvard [152]. More recent markets accommodate today's challenges in energy efficiency and heterogeneous hardware. Addressing energy efficiency, Chase et al. propose a market for processors and their electricity costs [30]. In this market, users bid according to their SLAs, and the datacenter operator allocates cycles to maximize welfare. Welfare from infrastructure-as-a-service is defined as revenue from user bids minus costs from datacenter expenses. The datacenter operator amortizes capital expenses over the server's lifespan and incurs operating expenses based on server power and electricity prices. Chase et al. are the first to consider markets for datacenter demand response, a framework to modulate computation based on electricity prices.

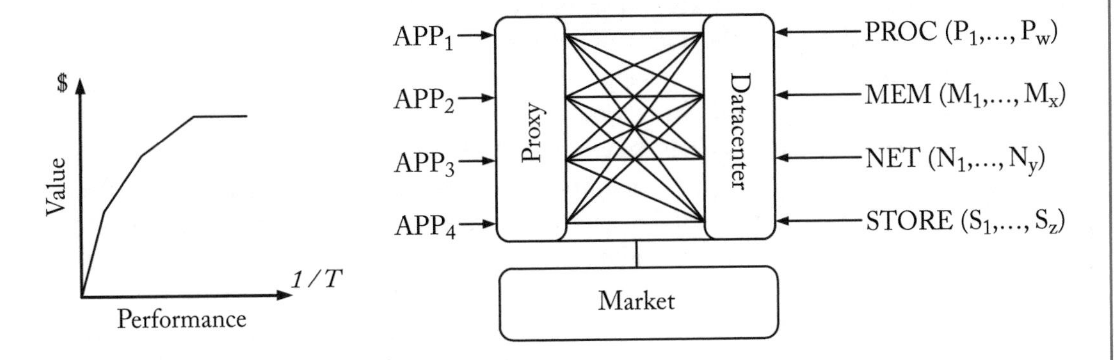

Figure 4.3: Service-level agreements (SLAs) specify user utility from performance. In a market mechanism for resource allocation, users and their applications interface with a proxy that bids for resources. The market allocates to maximize welfare, which is the sum of bids minus costs.

Addressing heterogeneous hardware, Guevara et al. propose a market for heterogeneous processor cores within a datacenter [58]. Users specify their value from performance with SLAs. Automated proxies profile task performance on varied hardware, analyze queueing dynamics, and link performance estimates to user SLAs to generate a bid. The market maximizes welfare, which is the sum of bids minus costs, by allocating bundles of heterogeneous processor cores based on task performance on the hardware types. A compute-intensive task would prefer big, high-performance cores whereas a memory-intensive task would be indifferent to small, low-power cores. Guevara et al. are the first to accommodate microarchitectural heterogeneity into a market mechanism, a step toward mitigating the management risks from mixing high-performance and low-power hardware.

In many markets for computer systems [30, 58], users state their true value, datacenters state their true costs, and the market allocates resources to maximize value minus cost. Unfortunately, in such markets, users can overpay. For example, a user might calculate his bid based on the marginal benefits to performance and pay his bid even when she is alone in the system. Wang and Martinez address this challenge with a price discovery mechanism [160]. Bidding agents automatically tune their bids so that they reflect demand for varied resources such as last-level cache capacity and power. The market allocates resources in proportion to bids to improve system throughput, out-performing a mechanism that divides resources equally.

4.5 RESOURCE ALLOCATION FOR FAIRNESS

Datacenters are built for different purposes and must be managed with different objectives. Datacenters built to supply infrastructure-as-a-service should allocate resources to optimize welfare by maximizing revenue and minimizing bids. Users get the hardware they pay for, which produces an allocation policy that might starve others of resources. Thus, a market mechanism is designed for performance and neglects the notion of fairness.

In contrast, datacenters built to supply shared capability should allocate resources according to some measure of fairness. The abstract notion of fairness translates into a number of desirable system outcomes. For example, a fair datacenter that allocates hardware to ensure equal progress among the many tasks that comprise a job would mitigate stragglers and reduce the critical path. In another example, a fair datacenter that allocates hardware to ensure that no user performs worse than under an equal share would incentivize sharing between strategic users. In both examples, datacenters ensure fairness but their definitions of fairness differ fundamentally—the former ensures fairness in performance outcomes whereas the latter ensures fairness in resource allocations. Recent research has made progress in pursuing both types of fairness.

Fair Outcomes. Fairness in performance outcomes is desirable in parallel or distributed systems in which many threads or tasks contribute computation on their pieces of the dataset before synchronizing. Ideally, computational load is distributed and resources are allocated to minimize the variance in task completion time. In practice, equalizing completion time is difficult for heterogeneous tasks and computer architects often pursue fairness by equalizing relative progress. Eyerman and Eeckhout survey system-level performance metrics, including fairness, for multi-programmed systems [44]. This survey includes a study by Gabor et al., which proposes the notion of equalizing relative progress for fairness [51, 52].

Suppose user i derives utility $u_i(x_i)$ when receiving allocation x_i in a shared system and derives utility $u_i(C)$ when receiving all resources C. User i's relative progress is defined as

$$U_i(x_i) = u_i(x_i)/u_i(C).$$

Note that we measure progress relative to performance when given complete system access (i.e., allocation C), but we could normalize performance under other scenarios (e.g., allocation X) as long as all users report performance relative to the same scenario. By using the same allocation scenario to normalize each user's performance, we obtain a standard definition of relative progress that is comparable across users. Finally, ensuring equal progress means finding allocations such that

$$\max_i U_i(x_i) \ / \ \min_j U_j(x_j) \rightarrow 1.$$

This allocation objective minimizes the gap in relative progress between user i and j who make the most and least progress, respectively, given their resource allocations.

Van Craeynest et al. pursue equal progress, in addition to other definitions of fairness, for multi-threaded and multi-programmed systems [157]. First, they measure relative progress. Second, they pursue equal progress when scheduling threads on a heterogeneous chip-multiprocessor with big and small cores. Assessing progress poses a practical challenge as the scheduler must predict and compare performance for a given program phase on both core types to determine relative progress. Estimating progress requires sampling performance under varied scenarios, examining historical performance under varied scenarios, or deriving analytical performance models. Performance impact estimation (PIE) is a method for accurately assessing performance under multiple allocation scenarios with big and small cores [158].

Principles for multi-threaded and multi-programmed architectures extend naturally to distributed systems and datacenters. At scale, datacenters compute for millions of tasks that are organized into thousands of jobs. Tails in the task latency distribution pose significant management challenges as straggling tasks lengthen a job's critical path [34]. Reducing the weight and length of the tail distribution is analogous to improving the progress of the slowest tasks in the system. For example, Zaharia et al. propose the LATE scheduler for Hadoop computation in heterogeneous systems [170]. The scheduler estimates the rate at which a task makes progress by examining the number of data records processed in a given time period. By extrapolating linearly, the scheduler estimates completion time. Then it speculatively duplicates tasks that have completion times farthest into the future and launches those duplicates on faster machines. In effect, the LATE scheduler pursues equal outcomes for tasks that comprise a job.

Although mechanisms for fair outcomes have their advantages in cooperative settings, they have several limitations in a competitive setting. When software tasks cooperate to complete a shared job, each task should request and receive whatever hardware resources are necessary to avoid contributing or lengthening the critical path. However, when software tasks are heterogeneous and compete for hardware resources independently, pursuing equal performance may require disproportionate allocations that harm some users and benefits others. For example, allocating memory to ensure equal progress for two tasks, one compute-intensive and the other memory-intensive, would divert the majority of memory to the second task to ensure its progress. When applied to heterogeneous tasks, mechanisms for fair outcomes risk equally poor outcomes for all tasks and are susceptible to strategic behavior in system settings with competitive users.

Fair Allocations. Fairness in resource allocations is desirable for shared and federated systems. Shared systems improve efficiency by co-locating multiple software tasks, increasing hardware utilization, and amortizing the platform's fixed power costs. For example, a shared system might arise when a web service provider builds a single system for search, mail, and word processing. In another example, a federated system might arise when multiple researchers pool their individual resources to create a larger computing cluster. In these and other settings, the system distributes hardware from a shared pool of resources to participating users. Whether users have an incentive to participate depends on mechanisms for fairness.

Strategic users who dislike allocations from the shared system might prefer private systems, even if they are less capable. For example, suppose Alice and Bob are researchers working on projects with different types of computational tasks. Each researcher has $100K to buy computers. Supposing Alice and Bob are strategic, they will ask themselves whether small, separate clusters or a large shared cluster will perform better. Supposing Alice and Bob share, they will ask themselves whether the allocation is fair and whether lying (i.e., misreporting hardware demands) is beneficial.

A game-theoretic perspective outlines desirable system properties in a system shared by strategic users. First, the system should provide sharing incentives (SI) such that each user performs at least as well as if she had an equal share of the resources. Second, the system should ensure envy-freeness (EF) such that each user prefers her own allocation over other users' allocations. Third, the system should ensure Pareto efficiency (PE) such that no other resource allocation would improve performance without harming someone. Finally, the system should provide strategy-proofness (SP) such that users who misreport their preferences for hardware cannot manipulate allocations. Researchers have devised allocation mechanisms that guarantee these desiderata for two system settings.

First, Dominant Resource Fairness (DRF) ensures game-theoretic desiderata for linear utility functions and complementary resources [54, 125]. Specifically, the Leontief utility function models task throughput in distributed systems, assuming that throughput increases linearly with allocation and that throughput is limited by the smallest allocation relative to user demands. In the equation below, Leontief utility u_L accounts for allocation x_r and user demand d for r types of resources.

$$u_L(x) = \min\left(\frac{x_1}{d_1}, \ldots, \frac{x_r}{d_r}\right).$$

The Loentief utility function accurately models user demand for complementary resources, such as processor cores and main memory capacity. For example, suppose user 1 requests {1 CPU, 4 GB} and user 2 requests {3 CPU, 1 GB}. Users 1 and 2 derive utility u_1 and u_2 from their processor allocation $x = (x_1, x_2)$ and memory allocation $y = (y_1, y_2)$.

$$u_1 = \min\left(\frac{x_1}{1}, \frac{y_1}{4}\right); \quad u_2 = \min\left(\frac{x_2}{3}, \frac{y_2}{1}\right).$$

DRF provides an allocation mechanism that generalizes max-min fairness to multiple resource types. For a single resource, a mechanism for max-min fairness maximizes the smallest allocation to any user. For multiple resources, however, a user may demand one resource more than others and heterogeneous users may prioritize resources differently. In this setting, DRF obtains each user's resource request, calculates resource shares, and identifies the dominant resource. The resource share is defined as the user's demand as a fraction system capacity. Among the user's demanded resources, the dominant resource is the one that corresponds to the largest resource share.

Returning to our example, user 1 requests {1 CPU, 4 GB} and user 2 requests {3 CPU, 1 GB}. Suppose the system has {9 CPUs, 18 GB} in total. Dividing requests by capacity, we find that users 1 and 2 request resource shares of {1/9, 2/9} and {6/18, 1/18}, respectively. User 1's dominant resource is memory capacity because its requested share of memory is 2/9, which is greater than its requested share of processors. Similarly, User 2's dominant resource is processor cores because its requested share of cores is 6/18, which is greater than its requested share of memory. Upon identifying these dominant resources, DRF maximizes allocations x and y subject to system capacity constraints and equal dominant shares. Allocating resources in this manner, DRF provably ensures game-theoretic desiderata—sharing incentives, envy-freeness, Pareto efficiency, and strategy-proofness.

$$\begin{array}{ccc} \max(x, y) & \text{subject to} & \\ x + 3y & \leq & 9 \\ 4x + y & \leq & 18 \\ \frac{2x}{9} & = & \frac{y}{3}. \end{array}$$

Second, Resource Elasticity Fairness (REF) ensures game-theoretic desiderata for nonlinear utility functions and substitutable resources [174, 175]. The Cobb-Douglas utility function models application performance by capturing diminishing marginal returns and substitution effects. In the equation below, Cobb-Douglas utility u_{CD} accounts for allocation x_r for r types of resources. The parameter α_r denotes the performance elasticity of resource r, which is determined by software behavior and its demands on hardware.

$$u_{CD}(x) = \prod_{r=1}^{R} x_r^{\alpha_r}.$$

The exponents in the Cobb-Douglas utility function model diminishing returns in computer architecture, such as Amdahl's Law as core count increases and limited locality as cache capacity increases. The product models substitution effects between resources—note that the change in utility when the allocation of resource i changes depends on allocations of other resources $-i$ (i.e., $\delta u/\delta x_i = f(x_{-i})$). This property of the Cobb-Douglas utility function is particularly important when modeling substitutable resources. For example, an application might substitute cache capacity for communication bandwidth without compromising utility. When users permit such substitution, the allocation mechanism has greater flexibility to pursue game-theoretic desiderata.

For example, consider last-level cache capacity and main memory bandwidth, substitutable resources. Suppose users 1 and 2 report the following utility functions:

$$u_1 = x_1^{0.6} y_1^{0.4}; \quad u_2 = x_2^{0.2} y_2^{0.8},$$

where x is the user's allocation of memory bandwidth and y is the user's allocation of cache capacity. User 1's performance elasticity of memory bandwidth is a bit larger than that of cache capacity, whereas user 2's performance elasticity of cache capacity is much greater than that of memory bandwidth. These utility functions are representative and fitted to the performance characteristics of canneal and freqmine from the PARSEC benchmark suite.

REF provides an allocation mechanism that guarantees game-theoretic desiderata. First, REF profiles the application, collecting performance data u for varied allocations $x = (x_1, \ldots, x_r)$. Second, REF fits the Cobb-Douglas utility function $u = \prod_{r=1}^{R} x_r^{\alpha_r}$ by applying a log transformation to produce $\log(u) = \sum_{r=1}^{R} \alpha_r \log(x_r)$ and performing a linear regression to find α_r. Third, REF scales users' elasticities such that they sum to one and are comparable across applications. Finally, REF allocates resources in proportions to elasticities.

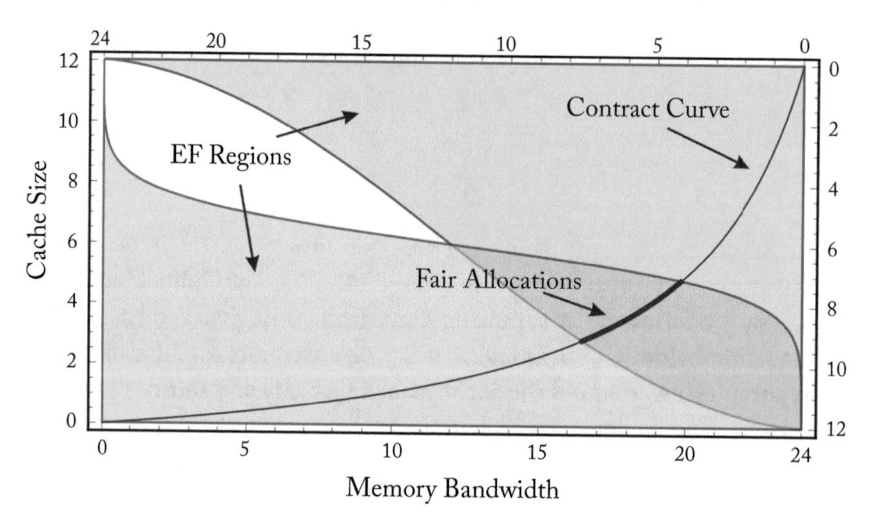

Figure 4.4: Fair allocations are envy-free and Pareto-efficient [174, 175]. The box illustrates the space of feasible allocations to two users who report utility functions $u_1 = x_1^{0.4} y_1^{0.6}$ and $u_2 = x_2^{0.2} y_2^{0.8}$. Fair allocations are those that are envy-free and Pareto-efficient (i.e., on the contract curve).

Returning to our example, suppose $u_1 = x_1^{0.6} y_1^{0.4}$ and $u_2 = x_2^{0.2} y_2^{0.8}$. Recall that x refers to a user's allocation of memory bandwidth. If the system supplies 24 GB/s of bandwidth, then user 1 should receive $24 \times 0.6/(0.6 + 0.2) = 18$ GB/s and user 2 should receive $24 \times 0.2/(0.6 + 0.2) =$

6 GB/s. Similarly, REF would divide 12 GB of last-level cache by assigning $12 \times 0.4/(0.4 + 0.8) = 4$ MB to user 1 and $12 \times 0.8/(0.4 + 0.8) = 8$ MB to user 2.

Figure 4.4 illustrates the space of feasible allocations and identifies the set of fair allocations. The Edgeworth box illustrates the possible divisions of 12 MB of last-level cache and 24 GB/s of memory bandwidth. The first user views allocations from the lower left corner and the second views them from the upper right such that one user receives an allocation and the other receives the remaining resources.

Fair allocations are envy-free and the Edgeworth box illustrates allocations for which the first user is envy-free (upper, blue region) and those for which the second user is envy-free (lower, red region). Fair allocations are also Pareto-efficient and the box illustrates efficient allocations with the contract curve. An allocation on this curve cannot be improved without harming a user's utility. Finally, fair allocations lie at the intersection of these three conditions. Zahedi and Lee prove that these allocations guarantee game-theoretic desiderata—sharing incentives, envy-freeness, Pareto efficiency, and strategy-proofness in large systems [174, 175].

CHAPTER 5

Hardware Simulation

Computer architects rely on simulators to tune sophisticated applications, organize system architectures, and evaluate emerging technologies. These research activities are particularly important for datacenter applications, given their qualitatively different demands on hardware. Yet simulating datacenter applications requires more software infrastructure and domain-specific knowledge than simulating conventional processor benchmarks. Indeed, this synthesis lecture starts with an application overview because deploying the workload, whether on a physical or simulated system, is often on the critical path for datacenter experiments. In this chapter, we assume the reader has identified and deployed a datacenter application for detailed study on a physical machine and we turn to workload simulation.

Computer architects must simulate datacenters at multiple scales. At the scale of a single server, architects rely on cycle-level simulators that track instructions as they progress through the processor and track data as they travel through the memory hierarchy. With such simulators, architects can assess the impact of an energy-efficient processor, a specialized functional unit, or an emerging memory technology. Simulators must obtain an instruction stream and invoke performance models—all while the application runs.

Datacenter applications pose several particular challenges. First, many applications require the full software system stack, including the operating system for network I/O and the Java virtual machine. Second, they require hefty initialization or pre-processing to populate main memory with input data, such as the web index or social network graph. Finally, they require simulation points that clearly account for domain-specific regions of interest, such as regions that start and end with web search query boundaries. With precisely targeted simulation, architects can evaluate varied hardware architectures and determine service time for individual computational tasks.

This synthesis lecture emphasizes a decoupled strategy in which the results of server-level simulators feed datacenter-level simulators. This chapter focuses on the server and the next focuses on the datacenter.

5.1 FULL SYSTEM SIMULATION

Computer architects have long relied on simulators and, historically, simulators have emphasized the processor. In the 1990s, architects focused on instruction-level parallelism and cache performance, which required extensible simulators to support a quantitative approach to research (e.g., Simplescalar [27]). In the 2000s, the focus shifted toward multiprocessors and new simulators that accurately modeled shared caches, the coherence protocol, and the on-chip network (e.g., GEMS [111]). This period also saw a broadening of computer architecture and its simulators, which looked beyond the processor to main memory (e.g., DRAMSim [140]) and looked beyond performance to power (e.g., Wattch [25]).

At present, computer architects are further broadening their field of study. Laterally, across component boundaries, they are studying new architectures for graphics processors that support general-purpose computing (e.g., GP-GPUSim [1]) and for domain-specific accelerators (e.g., Aladdin [144]). Vertically, across abstraction layers, they are evaluating emerging technologies (e.g., phase change memory [91, 133, 177]) and datacenter applications (e.g., the subject of this lecture). As the scope of study expands, so must simulator capability.

Demands on Simulators. Today's computer architects place a number of demands on system simulators, especially when studying datacenter applications. First, architects require full-system simulation. Datacenter workloads require the system software beneath an application, including an operating system, virtual machines, resource containers, libraries, and the network stack. The simulator supports this software infrastructure by mixing detailed performance models for specific hardware components (e.g., processor and memory system) with functional emulation for the rest (e.g., network stack). Beyond functional capability, emulation provides speed and supports sophisticated system checkpoints for precise and tractable simulation.

Second, architects require cycle-level timing models for components of interest. Models for the processor and memory are likely the most mature, given the degree of prior work in the former and the clearly defined timing parameters for the latter. Cycle-level models provide deep insight beyond today's hardware performance counters. Moreover, they provide extensible insight by accommodating new models, for new microarchitectures or emerging technologies.

Third, architects require simulators that support industry-strength instruction sets. Conventional wisdom says that x86 instruction set is particularly relevant for datacenter and server workloads whereas the ARM instruction set is well suited for mobile and embedded systems. Yet, to understand competitive forces and technological foundations, architects are increasingly interested in cross-over scenarios in which, for example, ARM processors deploy in datacenters [4]. As of this writing, the most widely used simulators support multiple instruction sets, but the depth of that support is uneven—MARSSx86 supports x86 particularly well [126] whereas gem5 is notable for its support for ARM [21].

This chapter describes simulation strategies for datacenter workloads using MARSSx86 [126], which provides many of the desiderata sought by architects. MARSSx86 provides full system simulation, which includes the full software stack and supports sophisticated datacenter workloads. MARSSx86 employs an emulation engine that extracts instruction streams to feed performance models. Moreover, the engine supports virtual machine and system checkpoints that amortize the cost of lengthy application warm up and permit precise simulation points, thereby improving tractability for long-running applications. Finally, as its name implies, MARSSx86 supports x86, which is the predominant architecture in server processors.

Although the author of this synthesis lecture and his research group use MARSSx86 and occasionally contribute to the community, they do not count themselves among core contributors. Rather, they are users who have surveyed a number of simulators and have chosen MARSSx86 for its merits. Readers who seek other simulators may consider gem5 [21], which is particularly good at modeling the memory system in multiprocessors and the ARM instruction set, or Flexus [47], which has been used by its developers to simulate a suite of datacenter workloads.

MARSSx86. Although today's computer architects can employ a rich set of capable simulators, datacenter research requires full-system simulation. MARSSx86 employs QEMU, an open-source hypervisor and machine emulator that supports a variety of instruction sets (e.g., x86, ARM) and the full software stack [121]. Figure 5.1 illustrates an instance of QEMU running on a host machine. A guest operating system and its applications run atop QEMU. QEMU runs programs and operating systems designed for the guest machine on a different host machine. It performs dynamic binary translation to convert guest computation to native computation on the host machine. The emulation engine translates processor computation from one instruction set to another and emulates the functionality of I/O devices.

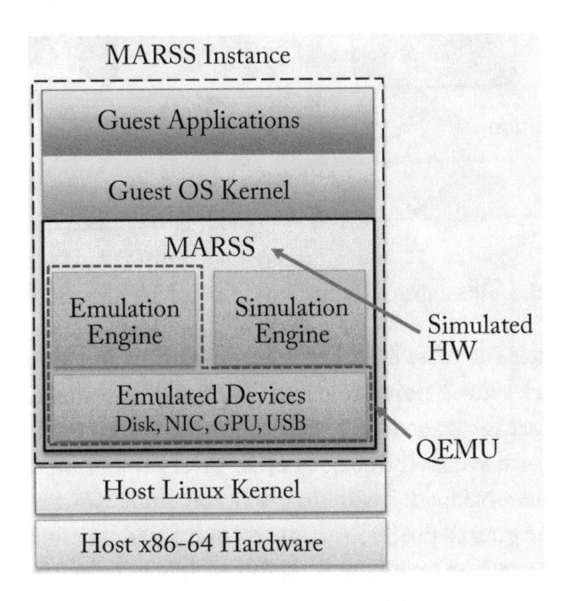

Figure 5.1: MARSSx86 and QEMU. Reproduced from MARSSx86 tutorial [128].

In Figure 5.1, MARSSx86 extends QEMU with a simulation engine, which employs the PTLsim performance model to simulate processor cores and caches [169], and employs the DRAMsim performance model to simulate main memory [140]. During guest computation, MARSSx86 and QEMU provide two paths to the host. The simulation path tracks both user and kernel instructions, supplies them to timing models for the processor, cache, and memory, and collects statistics. The emulation path supports system calls and I/O to a number of hardware devices (e.g., disk, network interface card, graphics processor).

Emulation vs. Simulation. Tightly coupled emulation and simulation provide a number of performance benefits. First, the cycle-level simulator and its timing models need not accommodate complex x86 opcodes, which are emulated by QEMU. Second, QEMU emulation permits the simulation of applications running an unmodified software stack. The operating system, user applications, and libraries (e.g,. pthreads) run in simulation mode to model hardware performance during computation of interest and run in emulation mode otherwise.

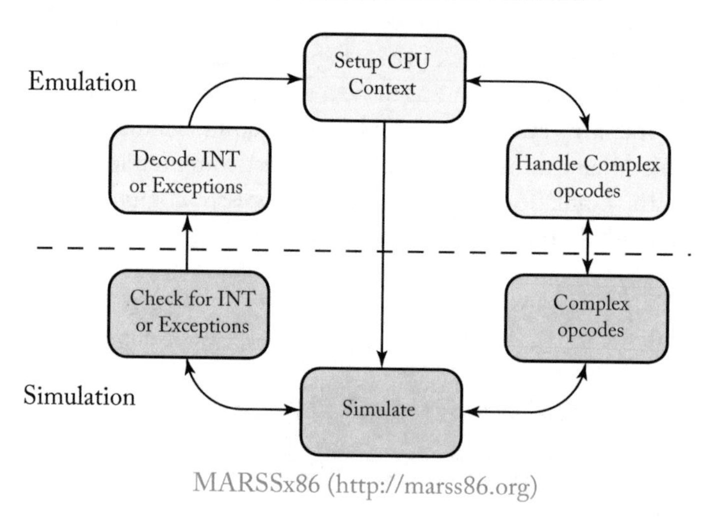

Figure 5.2: Emulation and Simulation. Reproduced from MARSSx86 webpage [127].

Figure 5.2 illustrates the close links between simulation and emulation. MARSSx86 begins in emulation mode, which sets up the system and the processor context. For example, a system that runs web search must start the operating system, start the search engine, and warm-up the query cache. Emulation is fast and avoids the costs of cycle-level simulation during lengthy initialization phases for the system and workload. Switching between emulation and simulation is inexpensive as both engines share the guest's processor contexts and physical memory.

When guest computation arrives at a region of interest, MARSSx86 switches to simulation mode. During simulation, MARSSx86 supplies instructions to cycle-level timing models and collects performance statistics. When simulation encounters complex x86 opcodes, MARSSx86 employs QEMU emulation to handle the opcodes and update the processor context before returning to simulation. When simulation encounters an exception, MARSSx86 employs QEMU emulation to decode the exception before returning to simulation.

Note that MARSSx86 can simulate kernel instructions from the exception handler. For example, consider a load instruction that misses in the translation lookaside buffer (TLB). MARSSx86 performs a detailed simulation of the page walk—which includes a number of memory transactions that request page table entries. If the page walk fails to produce a translation, the system encounters a page fault. MARSSx86 switches to emulation mode and switches the processor context to invoke the operating system's page fault handler. After the context switch, MARSSx86 returns to simulation mode to collect statistics as the operating system's page fault handler executes.

In simulation mode, MARSSx86 relies on two timing models. PTLsim provides cycle-level timing models for a variety of processor components—cores, caches, networks-on-chip. MARSSx86's user builds the processor timing model by connecting the timing models for varied components. For example, the user might employ out-of-order, write-back, write-through, or coherent caches for each level in the hierarchy, and a bus for the network-on-chip to build a virtual machine for simulation. Of course, the user could also implement her own timing models for any of these components.

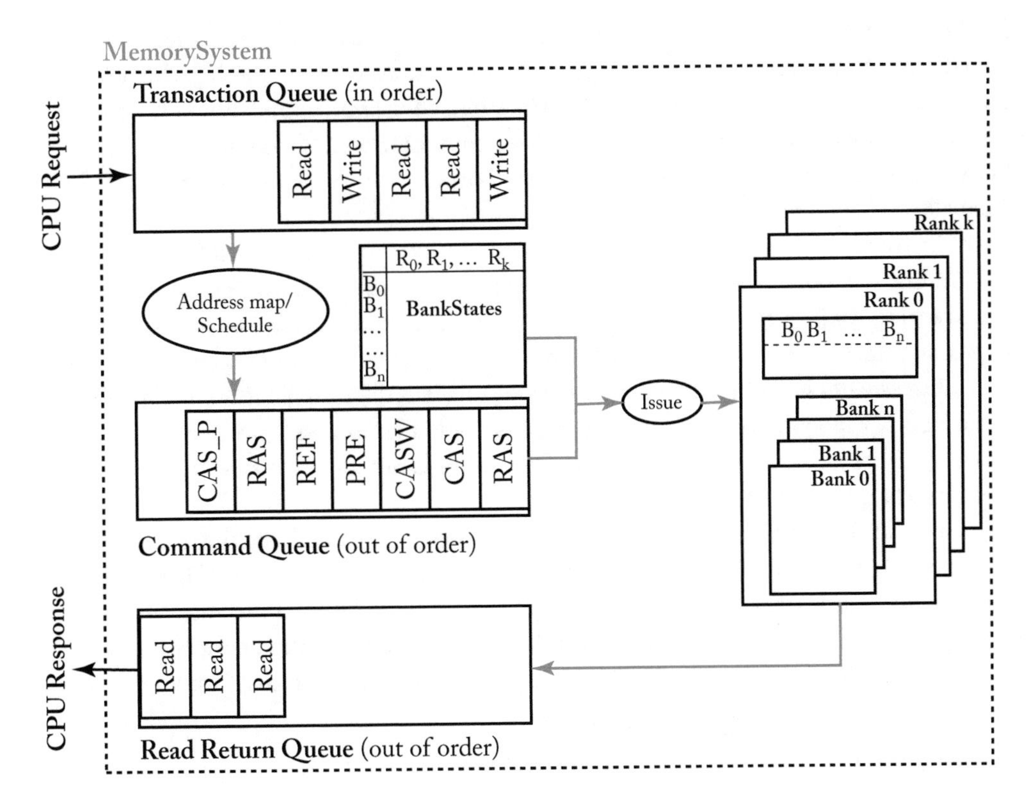

Figure 5.3: Main memory system as modeled by DRAMSim2. Reproduced from [140].

PTLsim sends last-level cache misses to DRAMSim, which provides cycle-level timing models for main memory. DRAMSim specifies and enforces the timing parameters that govern memory transaction scheduling, which is illustrated in Figure 5.3. For example, the memory controller must wait some number of cycles after activating a DRAM row and before reading a column of data from it. DRAMSim2 also estimates main memory power by specifying current draw during varied modes of operation, simulating the time spent in each mode, and multiplying the weighted average for current and the operating voltage.

Beyond these timing and power models, MARSSx86 is extensible and can be integrated with a variety of other models. For example, McPAT models can use PTLsim statistics to estimate activity factors and power dissipated by microarchitectural components [96]. Although MARSSx86 emulates many devices, such as the graphics processor and network interface, it could simulate these devices if provided performance models, like those in GPGPUsim [1] or gem5 [21].

5.2 INITIALIZING THE SYSTEM

MARSSx86 provides light-weight mechanisms to switch between emulation and simulation. Emulation is fast whereas simulation provides microarchitectural detail. Judiciously switching between modes provides performance and tractability for large applications. A computer architect can emulate the majority of an application's computation and simulate only specific regions of interest. Moreover, MARSSx86 provides mechanisms for checkpointing and fast-forwarding, which allows the architect to initialize system state once, checkpoint immediately before the region of interest, and simulate a targeted application phase with reproducible measurements.

In this section, we describe the steps for initializing the simulator and simulating concisely defined regions of interest within an application. We use representative code snippets to illustrate the process; the code is not intended to stand alone and provide comprehensive documentation.

Disk Image. Full-system simulation requires a lengthy initialization phase—the operating system boots, the application starts, the data is loaded from disk into main memory, and the processor caches warm up. MARSSx86 provides a simple initialization flow that exploits QEMU's capabilities as a hypervisor.

Specifically, initialization requires creating a disk image and installing an operating system onto that image. In the example below, QEMU commands create a 10GB disk image in the qcow2 format and installs the operating system by booting a QEMU instance with 4GB of main memory from an ISO image of an optical disk. After installing the operating system onto the disk image, we can re-launch the QEMU instance to boot the operating system and arrive at the command line for the guest operating system.

```
> qemu-img create -f qcow2 imageName.qcow2 10G

> qemu-system-x86 -m 4G
    -drive file=imageName.qcow2,cache=unsafe
    -cdrom mini.iso -boot d -k en-us

> qemu-system-x86 -m 4G
```

```
    -drive file=imageName.qcow2,cache=unsafe
    -k en-us -nographic
```

The steps above prepare a disk image with a complete system software stack. But launching a QEMU instance with the above commands provides only an emulation path from the guest application and operating system, through the QEMU hypervisor and emulation engine to the host system.

Model Configuration. Simulation requires specifying parameters for hardware components. MARSSx86 specifies processor parameters by defining datapath resources, cache organization, and coherence protocol. The simulator provides default models for out-of-order and in-order cores and parameterizes key datapath features (e.g., issue width, register file size, etc.). It provides default models for caches and parameterizes cache geometries (e.g., size and associativity).

Finally, it models memory latency with either a simple model or the sophisticated, cycle-level DRAMSim model [140]. The DRAMSim model defines the DRAM organization—e.g., number of ranks, banks, and rows—and the DRAM timing parameters. Parameterized timing parameters are particularly helpful when simulating varied memory interfaces and technologies. For example, LP-DDR3 and DDR4 interfaces differ in transition times between power modes; DDR4 requires more time to wake up from sleep modes as its delay-locked loops must re-synchronize. DRAM models also provide a baseline and a framework for evaluating emerging technologies, which present new timing values and constraints.

After configuring processor and system parameters, we compile and create a new instance of the MARSSx86 simulator. We first compile DRAMSim to create a shared library to be used by MARSSx86. We then build MARSSx86, specifying the processor configuration file and the path to the memory simulator. The build produces a new QEMU binary that integrates PTLSim. Finally, we create a simulator configuration file (simcfg) to specify the machine and memory system configuration files, the destination files for logs and statistics, and other settings when running the simulator.

```
> cd <path-to-marssx86>

> make libdramsim.so

> scons -Q config=<path-to-proc-config> dramsim=<path-to-dramsim>

> ./qemu/qemu-system-x86_64 -m 4G
    -drive file=imageName.qcow2,cache=unsafe -nographic
    -simconfig <path-to-simcfg>
```

Note that the steps above launch an instance of the MARSSx86 simulator by launching an instance of QEMU with an initialized disk image and specifying the simulator configuration.

After the operating system boots, the computer architect can perform arbitrary computation on the full system software stack within the guest system.

5.3 INVOKING THE SIMULATOR

The MARSSx86 instance will either simulate the computation by invoking PTLsim and DRAM-sim, as configured, or emulate the computation by invoking QEMU's engine and models for complex opcodes and system devices. The computer architect must start the simulator before a region of interest and kill the simulator to produce model outputs and statistics. MARSSx86 provides a number of PTLCalls that provide an interface between QEMU and PTLsim. These PTLCalls determine when and how MARSSx86 invokes PTLSim.

- `ptlcall_switch_to_sim()`: Leave emulation mode and enter simulation mode
- `ptlcall_switch_to_native()`: Leave simulation mode and enter emulation mode
- `ptlcall_kill()`: Terminate the simulation
- `ptlcall_checkpoint_and_shutdown(chkpt_name)`: Take a snapshot of the virtual machine, save it as chkpt_name, and shut down the virtual machine.

To simulate computation launched from the command line, we create binaries that wrap around the basic PTLCalls. For example, we define `start_sim` to switch into simulation mode. Similarly, we define `stop_sim` with `ptlcall_switch_to_native` call and `kill_sim` with `ptlcall_kill`.

```
// start_sim.c
#include <stdlib.h>
#include <stdio.h>
#include "ptlcalls.h"

int main(int argc, char* argv[]) {
     printf("Starting simulation\n");
     ptlcall_switch_to_sim();
     return;
}
```

After compiling these utilities that invoke PTLCalls, we can start simulation, invoke computation, and end simulation from the command line. We denote the guest machine's command line as `$` to distinguish it from the host machine's command line, which is denoted as `>`. To simulate an application called `helloWorld`, we invoke PTLCalls as follows.

```
$ ./start_sim;
$ ./helloWorld;
$ ./stop_sim;
```

PTLCalls provide the requisite functionality for simulating datacenter applications that require the full system software stack. However, relying on simulation from the guest machine's command line is unwieldy, especially for datacenter applications that require lengthy initialization and warm up. To simulate more efficiently and with greater precision, we rely on checkpoints and embed PTLCalls within the application itself.

5.4 CREATING CHECKPOINTS

QEMU provides sophisticated mechanisms for system checkpoints, which are snapshots of the qcow2 disk image. These snapshots save the state of the virtual machine at a particular point in time. Snapshots are particularly helpful when the architect wishes to initialize a complex application and then execute a number of diverse tasks. For example, suppose the architect wishes to study web search. She can initialize the search engine in emulation mode, create a checkpoint, then execute each query of interest without repeating the initialization.

MARSSx86 supports multiple approaches to creating checkpoints. First, we can wrap the corresponding PTLCall (`ptlcall_checkpoint_and_shutdown(chkpt_name)`) in an executable utility and invoke it from the command line. Alternatively, we can embed the PTLCall into an application's source code. Note that checkpoints are specific to the system configuration (e.g., the number of cores, datapath specification, and cache organization). The code below for helloWorld creates a checkpoint, which is named according to an environment variable, immediately before printing the string.

```
// helloWorld.c
#include <stdlib.h>
#include <stdio.h>
#include "ptlcalls.h"

int main(int argc, char* argv[]) {
     char* chkpt_name = getenv("CHECKPOINT_NAME");
     if(chkpt_name != NULL) {
          printf("Creating checkpoint with name %s\n", chkpt_name);
          ptlcall_checkpoint_and_shutdown(chkpt_name);
     }
     printf("Hello World\n");
     ptlcall_kill();
     return;
}
```

Prior to running the program, set the environment variable to define the checkpoint name. Afterward, we can use the `qemu-img` utility to obtain information on previously created checkpoints or delete specific checkpoints.

```
$ export CHECKPOINT_NAME=helloWorldChkpt
$ ./helloWorld

$ ./qemu/qemu-img info imageName.qcow2
$ ./qemu/qemu-img snapshot -d helloWorldChkpt imageName.qcow2
```

To load a virtual machine from a checkpoint, we add the "`-loadvm chkpt_name`" when launching a new MARSSx86 instance. See the command line below.

```
> ./qemu/qemu-system-x86_64 -m 4G
    -drive file=imageName.qcow2,cache=unsafe -nographic
    -simconfig <path-to-simcfg>
    -loadvm helloWorldChkpt -snapshot
```

In summary, the QEMU hypervisor provides sophisticated mechanisms for taking snapshots of virtual machines. These snapshots provide complete system checkpoints, which are invaluable when simulating datacenter applications. Checkpoints provide a mechanism to initialize system and application state, permitting simulations for precisely defined regions of interest. Checkpoints are created in emulation mode, which means that initialization is fast. Checkpoints should be used in conjunction with fast-forwarding in simulation mode, which invokes PTLsim timing models, warms up simulation state, and provides statistics that more accurately represent processor and memory behavior.

Full-system simulation with virtual machine management and sophisticated timing models permit new and precise studies for datacenter workloads. We can deploy complex software frameworks (e.g., Apache Hadoop for MapReduce) or platforms that require the Java virtual machine (e.g., Apache Nutch for web search). Moreover, we can manage the big data sets required to exercise these applications by staging data into the virtual machine and taking system checkpoints. In this section, we illustrate these capabilities for three representative datacenter workloads—Spark, GraphLab, and Nutch web search.

5.5 CASE STUDY: SIMULATING SPARK

Apache Spark is a fast and general engine for large-scale data processing [172, 173]. The computational framework has attracted interest for its efficiency on general execution graphs (i.e., beyond MapReduce) and its judicious use of main memory for performance. In addition to its run-time system and task scheduler, Spark provides a unified platform for several programming models—GraphX for graph processing, MLlib for machine learning, Spark Streaming for real-time data processing, and Spark SQL for data stores.

Spark's unified platform leads to smaller run-time system (as measured in code size) and greater engine performance. Moreover, Spark programmers can compose functionality from multiple computational models. For example, a user could query a data store with Spark SQL and then feed query results into PageRank computation with Spark GraphX. Because the Spark engine implements mechanisms that cache frequently accessed data structures in main memory, data flows from one computational model to another without expensive transfers through the distributed file system.

Spark's use of main memory is particularly important for iterative, multi-pass computation, which is typical in machine learning and graph analytics. For such workloads, conventional Hadoop MapReduce is slow as every computational iteration reads data from and writes data to the distributed file system. Network and disk I/O affect performance in the loop body. In contrast, Spark is fast as every computational iteration reads data from and writes data to main memory. Spark's performance is one to two orders of magnitude greater than Hadoop's for iterative computation.

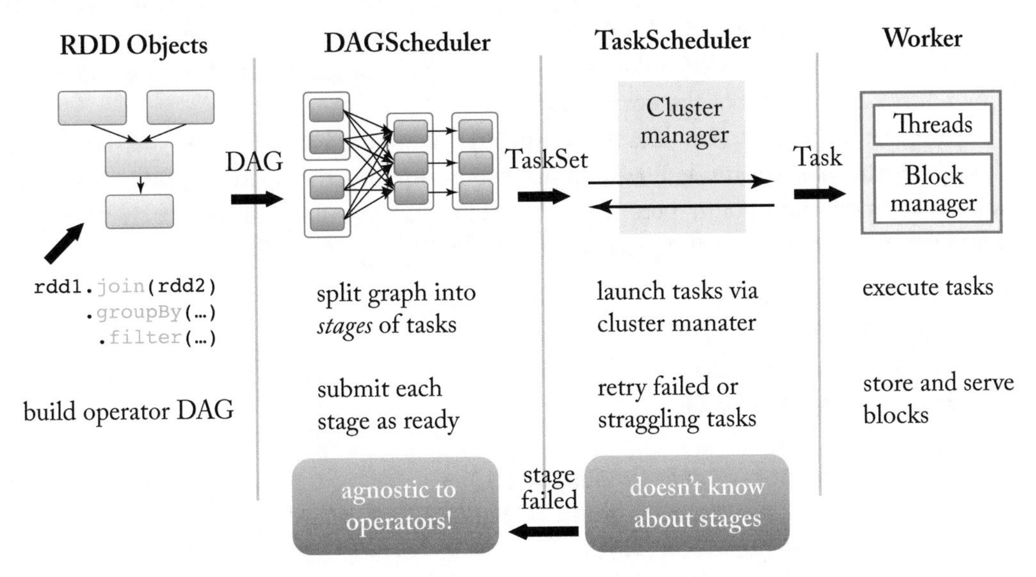

Figure 5.4: Spark run-time system codifies dependences between operators, creates tasks, and distributes them to workers. Reproduced from talk on Spark internals [171].

The basic Spark data structure is the resilient distributed dataset (RDD), which is created from data retrieved from external storage (e.g., distributed file system) or produced from a transformation on another RDD. A Spark program is a series of transformations and actions on RDDs. As shown in Figure 5.4, the Spark engine analyzes the program to construct a directed acyclic graph (DAG) that captures dependencies between RDD operators, splits operator computation into tasks, and submits tasks to the cluster manager and its workers as soon as their dependencies are satisfied.

```
wordCount = textFile.
    flatMap(lambda line: line.split()).
```

```
        map(lambda word: (word, 1)).
        reduceByKey(lambda a, b: a + b);

wordCount.collect();
```

The code above illustrates an implementation of `wordCount`. The `textFile` RDD has been retrieved from the data store and the program performs a series of three transformations. Each transformation specifies a function defined by lambda abstractions—function definitions that are not explicitly bound to an identifier. First, the `flatMap` transformation defines a function that takes the argument `line` and splits the line into words. This function is applied to each line in `textFile` linearized with the `flatMap` transformation. Second, the `map` transformation defines a function that takes a word and produces a key-value pair. This function is applied to each word produced by `flatMap`. Finally, the `reduceByKey` transformation defines a function for summation and applies it to key-value pairs produced by `map`, grouped by key. The final `collect` action returns all elements of the `wordCount` RDD.

Suppose we wish to simulate the wordCount computation and focus specifically on the reduceByKey transformation. Full-system simulation supports the Spark engine and its software stack, and system checkpoints permit precise simulation points. First, we create a utility that invokes the PTLCalls for creating checkpoints and stopping simulation.

```
// ptlcalls.cpp
#include <stdlib.h>
#include <iostream>
#include "ptlcalls.h"

extern "C" void create_checkpoint() {
    char* chkpt_name = getenv("CHECKPOINT_NAME");
    if(chkpt_name != NULL) {
        printf("Creating checkpoint with name %s\n", chkpt_name);
        ptlcall_checkpoint_and_shutdown(chkpt_name);
    }
}

extern "C" void stop_simulation() {
    printf("Stopping simulation\n");
    ptlcall_kill();
}
```

We compile the utility and create a shared library that will be invoked by the Python program that implements wordCount.

```
$ g++ -c -fPIC ptlcalls.cpp -o ptlcalls.o
```

```
$ g++ -shared -Wl,-soname,libptlcalls.so -o libptlcalls.so ptlcalls.o
```

Next, we include the library into wordCount's Python implementation. Note that we split the transformations on the text lines so that the first two transformations are applied and computed before creating the checkpoint. Because transformations are defined on an RDD and computed lazily (i.e., only when needed to complete an action), we must perform the `collect` action before the checkpoint. The code below precisely defines the region of interest for simulation by de-limiting the region with a `create_checkpoint` at the beginning and a `stop_simulation` at the end.

```
// wordcount.py
from ctypes import cdll lib =
cdll.LoadLibrary('./libptlcalls.so')

[....]

counts = lines.
     flatMap(lambda x: x.split(' ')).
     map(lambda x: (x, 1))
output = counts.collect()
lib.create_checkpoint()

counts = counts.reduceByKey(sample)
output = counts.collect()
lib.stop_simulation()
```

We create the PTLCall utilities and modify the wordCount code for computation on the guest (i.e., within a MARSSX86/QEMU instance). In practice, a computer architect boots the guest operating system to develop and compile code within the guest. At the end of this process, the guest's disk image should contain the PTLCall utilities and the modified applications. Finally, to create the checkpoint, the architect launches a MARSSx86/QEMU instance with the updated disk image. Within this guest machine, the architect defines a checkpoint name and runs the modified application that invokes the `create_checkpoint` library. The `qemu-img` utility will confirm the creation of a new checkpoint within the designated disk image. Note that this process creates the checkpoint but does not invoke the simulator.

```
> ./qemu/qemu-system-x86_64 -m 4G
     -drive file=imageName.qcow2,cache=unsafe -nographic

$ export CHECKPOINT_NAME=wordCountChkpt
$ cd <path-to-spark-binary>
$ ./spark-submit wordcount.py textFile
```

```
> ./qemu/qemu-img info imageName.qcow2
```

We simulate from the checkpoint by configuring the processor and memory system's parameter files, and by configuring the simulator parameter file as shown below. The example configures the simulator by specifying the output files, the number of cycles to wait before logging statistics (e.g., 10 M), and the processor and memory parameter files.

```
// wordCount.simgcfg
- logfile wordCount.log
- stats wordCount.yml
- startlog 10M
- machine single_core
- corefreq 3G
- dramsim-device-ini-file DDR3_micron_32M_8B_x4_sg125.ini
- dramsim-results-dir-name wordCount
- run
- kill-after-run
```

Given a simulator configuration file, we launch a new instance of MARSSx86/QEMU with the virtual machine checkpoint that was created for the disk image. Note that the checkpoint name corresponds to one of the snapshots associated with the disk image.

```
> ./qemu/qemu-system-x86_64 -m 4G
    -drive file=imageName.qcow2,cache=unsafe -nographic
    -simconfig wordCount.simgcfg
    -loadvm wordCountChkpt -snapshot
```

5.6 CASE STUDY: SIMULATING GRAPHLAB

GraphLab is a framework for large-scale graph analytics, which are often characterized by sparse data dependencies, local computation, and iterative updates [105, 106]. Examples of such computation include clustering, PageRank, pattern mining, reachability, and shortest-path analysis. Because these kernels often analyze large and sparse graphs, as illustrated in Figure 5.5, their computational behaviors exhibit poor locality, limited data parallelism, limited scalability, and I/O bottlenecks. Conventional MapReduce frameworks perform poorly for parallel graph analysis as the iterative nature of the computation requires re-loading and re-processing the graph; intermediate results are written to the distributed file system.

GraphLab presents an alternative approach for iterative graph analysis. It captures and enforces data dependences between graph vertices, but updates the data associated with each vertex asynchronously in a way that permits high-performance, parallel execution models. Specifically, the GraphLab framework provides a programming model in which the user (i) represents data

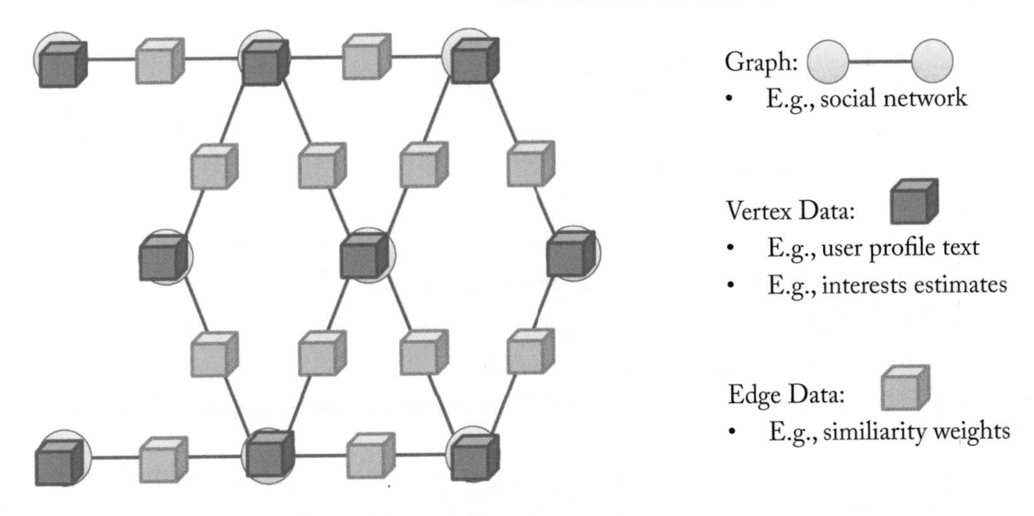

Figure 5.5: Graph structure illustrates the role of vertices and edges, both of which may be associated with data. Reproduced from talk on large-scale machine learning on graphs [57].

as a graph, (ii) specifies update functions on the graph, (iii) specifies a consistency model, and (iv) specifies a task scheduling policy. The run-time system pursues parallelism to achieve the specified updates subject to constraints imposed by the consistency model and scheduling policy.

In the first GraphLab example, we will create a recommendation system that is trained by users who have assigned ratings to movies they have watched. Toward this goal, we launch a MARSSx86/QEMU instance and, within the guest machine, obtain a dataset with movie ratings.

```
$ wget files.grouplens.org/datasets/movielens/ml-10m.zip
$ unzip ml-10m.zip
$ sed 's/::/,/g' ml-10M100K/ratings.dat > ratings.csv
```

Next, we employ GraphLab libraries to create the recommendation system with the dataset. The `gl.recommender.create` library chooses a recommendation model, factorization recommender in this case, based on the input dataset's format. The programmer can choose other types of recommendation models (e.g., item similarity, ranking factorization, popularity-based). She can also explicitly specify model structure (e.g., number of latent factors, maximum number of iterations during training, etc.).

```
//recommender.py
import graphlab as gl
data = gl.SFrame.read_csv(
     'ratings.csv',
     column_type_hints = {'rating':int},
     header=TRUE
```

```
)
model = gl.recommender.create(
      data,
      user_id='userid',
      item_id='movieid',
      target='rating'
)
results = model.recommend(users=None, k=5)
print results
```

The trained model can recommend new movies. The `model.recommend` function returns the k-highest scored items for each user. Counter-intuitively, specifying `users=None` returns recommendations for all users.

Suppose we wish to simulate the computation required during model creation and training. As before, she creates a utility that invokes the PTLCalls for creating checkpoints and stopping simulation.

```
// ptlcalls.cpp
#include <stdlib.h>
#include <iostream>
#include "ptlcalls.h"

extern "C" void create_checkpoint() {
      char* chkpt_name = getenv("CHECKPOINT_NAME");
      if(chkpt_name != NULL) {
            printf("Creating checkpoint with name %s\n", chkpt_name);
            ptlcall_checkpoint_and_shutdown(chkpt_name);
      }
}

extern "C" void stop_simulation() {
      printf("Stopping simulation\n");
      ptlcall_kill();
}
```

We compile the utility and create a shared library that will be invoked by the Python program that implements the graph analysis for model training and prediction.

```
$ g++ -c -fPIC ptlcalls.cpp -o ptlcalls.o
$ g++ -shared -Wl,-soname,libptlcalls.so -o libptlcalls.so ptlcalls.o
```

Next, we include the library in the recommender's Python implementation. The code below precisely defines the region of interest for simulation by de-limiting the region with a create_checkpoint at the beginning and a stop_simulation at the end. In this example, we simulate model training; we could repeat the same process to simulate model evaluation.

```
// wordcount.py
from ctypes import cdll lib =
cdll.LoadLibrary('./libptlcalls.so')

[....]

lib.create_checkpoint()
model = gl.recommender.create(
     data,
     user_id='userid',
     item_id='movieid',
     target='rating'
)
lib.stop_simulation()
results = model.recommend(users=None, k=5)
print results
```

We launch an instance of MARSSx86/QEMU, define an environment variable with the checkpoint name, and run the recommendation system to create the checkpoint.

```
> ./qemu/qemu-system-x86_64 -m 4G
     -drive file=imageName.qcow2,cache=unsafe -nographic

$ export CHECKPOINT_NAME=recommendChkpt
$ cd <path-to-recommend-py>
$ ./python recommender.py

> qemu-img info imageName.qcow2
```

Given a simulator configuration file, we launch a new instance of MARSSx86/QEMU with the virtual machine checkpoint that was created for the disk image. The resulting cycle-level simulation will collect statistics for the region of interest only, which is the GraphLab library that creates a recommendation model.

```
> ./qemu/qemu-system-x86_64 -m 4G
     -drive file=imageName.qcow2,cache=unsafe -nographic
     -simconfig graphlab.simgcfg
     -loadvm recommendChkpt -snapshot
```

In a second GraphLab example, we will perform Kmeans clustering or airlines and flight information for 2008. First, we obtain the dataset.

```
$ wget stat-computing.org/dataexpo/2009/2008.csv.bz2
$ bzip2 -d 2008.csv.bz2
```

Next, we implement the clustering algorithm with GraphLab libraries, import the PTL-Call libraries that we built, and de-limit the region of interest—Kmeans model construction, in this case—with a `create_checkpoint` at the beginning and a `stop_simulation` at the end.

```
// clustering.py
from ctypes import cdll
lib = cdll.LoadLibrary('./libptlcalls.so')

import graphlab as gl
from math import sqrt
data_url = '2008.csv'
data = gl.SFrame.read_csv(data_url)
data_good, data_bad = data.dropna_split()
n = len(data_good)
k = int (sqrt( n / 2.0))

print "Starting k-means with %d clusters" %k
lib.create_checkpoint()
model = gl.kmeans.create(data_good, num_clusters=k)
lib.stop_simulation()

model['cluster_info'][['cluster_id','__within_distance__', '__size__']]
```

After instrumenting the Python code, we run kmeans clustering within an instance of MARSSx86/QEMU. Doing so saves the system snapshot to the disk image for simulation. The simulation itself requires launching a new instance of MARSSx86/QEMU with the checkpoint.

```
> ./qemu/qemu-system-x86_64 -m 4G
    -drive file=imageName.qcow2,cache=unsafe -nographic

$ export CHECKPOINT_NAME=kmeansChkpt
$ cd <path-to-kmeans-py>
$ ./python kmeans.py

> ./qemu/qemu-img info imageName.qcow2

> ./qemu/qemu-system-x86_64 -m 4G
    -drive file=imageName.qcow2,cache=unsafe -nographic
    -simconfig graphlab.simgcfg
```

```
    -loadvm kmeansChkpt -snapshot
```

5.7 CASE STUDY: SIMULATING SEARCH

Apache Solr provides an open-source framework for search with diverse queries on a large, real-world index [9, 11]. Search requires strict service quality on a scalable system. Queries enter the system via an aggregator, which distributes queries to index serving nodes. Each node ranks the pages and the ranker returns captions to the aggregator. The search queries are important determinants of workload behavior. Queries differ in the number of query terms, the logical connectors of those terms, and the sophisticated operators (e.g., wildcards) that increase the relevance of the results [23]. Query complexity correlates strongly with response time.

We deploy Apache Solr on a single index serving node. Solr is an open-source, well-documented, and configurable search engine. It supports full-text search, permits user-extensible caching, and provides logs with server statistics. SolrCloud supports distributed search to accommodate high query volumes. It partitions the index, distributes it across multiple nodes, and serves queries at higher rates. In simulation, we study Solr on a single node with the expectation that findings on one node will generalize to Solr multiple nodes since (i) the index can be partitioned or replicated across nodes such that each node's dataset is representative of the broader dataset; (ii) every query is distributed to each Solr node; and (iii) query parallelism requires no communication between nodes and only modest communication between nodes and the aggregator.

Simulating an index serving node permits a number of studies. We might design new processor microarchitectures and memory systems for query processing. We might deploy new machine learning algorithms to improve the relevance of page rank. We might analyze query complexity and its effect on end-to-end system behavior. And we might devise new management strategies for managing resources and scheduling tasks. Each of these research directions benefit from the insight and precision of cycle-level simulation.

First, we illustrate Apache Solr to search an example dataset, which describes the inventory in a warehouse. We download and install the necessary frameworks: Java, Solr, and cURL. The solr distribution provides a number of binaries that start the search engine (`start.jar`) and index a dataset (`post.jar`). Suppose the warehouse inventory is represented by a series of XML documents each of which describes an item with a number of fields. For example, the following describes a computer monitor.

```
// monitor.xml
<add><doc>
<field name="id">3007WFP</field>
<field name="name">Dell Widescreen UltraSharp 3007WFP</field>
<field name="man">Dell, Inc.</field>
<field name="cat">electronics and computer</field>
<field name="features">30" TFT active matrix LCD, 2560 x 1600</field>
<field name="weight">401.6</field>
```

```
<field name="price">2199</field>
<field name="inStock">true</field>
</doc></add>
```

We invoke `post.sh` to index the inventory, either for one or all the XML documents, within an instance of MARSSx86/QEMU.

```
$ ./post.sh monitor.xml
$ ./post.sh *.xml
```

To execute queries against the indexed data, we start the search engine and issue queries to the engine via its network address and port. We launch the search engine with `start.jar`.

```
$ java -jar start.jar &
$ curl "http://localhost:8983/solr/collection1/
     select?q=inStock:false&wt=json&wt=json&fl=id,name&indent=true"
```

cURL is a command-line tool for transferring data between protocols and is the mechanism for issuing a query to the search engine. We refer the reader to a specific tutorial for Solr Query Syntax.[1] In this example, we employ a few representative query features. The collection specifies the set of indexed documents and a select operator that returns items for which the `inStock` field is false. The search engine returns results in the `json` format and reports values from two fields, `id` and `name`, for every result. Finally, the `indent` formats the results to produce the following output.

```
s=4 status=0 QTime=10
{
   "responseHeader":{
      "status":0,
      "QTime":10,
      "params":{
         "fl":"id,name",
         "indent":"true",
         "q":"inStock:false",
         "wt":"json"}},
   "response":{"numFound":4,"start":0,"docs":[
      {
         "id":"F8V7067-APL-KIT",
         "name":"Belkin Mobile Power Cord for iPod w/ Dock"},
      {
         "id":"IW-02",
```

[1] www.solrtutorial.com/solr-query-syntax.html

```
        "name":"iPod & iPod Mini USB 2.0 Cable"},
      {
        "id":"EN7800GTX/2DHTV/256M",
        "name":"ASUS Extreme N7800GTX/2DHTV (256 MB)"},
      {
        "id":"100-435805",
        "name":"ATI Radeon X1900 XTX 512 MB PCIE Video Card"}]
  }
}
```

Note that we are issuing queries at the command line, which means we should begin and end the simulation at the command line as well. Specifically, we create utilities that invoke PTL-Calls for creating checkpoints and stopping the simulation.

```
// create_checkpoint.c
#include <stdio.h>
#include "ptlcalls.h"

int main(int argc, char* argv[]) {
     if (argc > 1) {
          char* chkpt_name = argv[1];
          printf("Creating checkpoint %s\n", chkpt_name);
          ptlcall_checkpoint_and_shutdown(chkpt_name);
     } else {
          printf("No checkpoint name provided\n");
     }
     return;
}


// stop_sim.c
#include <stdio.h>
#include "ptlcalls.h"

int main(int argc, char* argv[]) {
     printf("Stopping simulation\n");
     ptlcall_switch_to_native();
     return;
}
```

With these utilities, we can precisely define a simulation for a query at the command line. Note that we issue a single query for this checkpoint. However, we could just as easily write a shell script with a sequence of cURL commands that correspond to a sequence of queries. The shell script would replace the individual cURL command in this example to create a more sophisticated benchmark. Thus, we can employ diverse strategies to create and simulate search benchmarks with varied query length, complexity, and content.

```
$ create_checkpoint queryChkpt;
$ curl "http://localhost:8983/solr/collection1/
    select?q=inStock:false&wt=json&wt=json&fl=id,name&indent=true"
$ stop_sim
```

Finally, we load a virtual machine from the checkpoint and launch an instance of MARSSx86/QEMU with the desired simulator configurations.

```
> ./qemu/qemu-system-x86_64 -m 4G
    -drive file=imageName.qcow2,cache=unsafe -nographic
    -simconfig <path-to-simcfg>
    -loadvm queryChkpt -snapshot
```

Beyond the simple example for warehouse inventory, we wish to index diverse documents that span many topics. Apache Solr indexes from data files or crawled websites. Apache Nutch is an open source web crawler that can, for example, crawl Wikipedia to produce a document corpus for Solr [11]. By configuring Nutch, we can tune the size of the index and the breadth of the crawl. The user can employ Nutch to crawl and index a variety of networks.

If we wish to simulate queries against a Wikipedia index, we can forgo the crawl and rely on pre-assembled datasets from Wikimedia. These datasets provide the Wikipedia document corpus in XML format. For example, we obtain and decompress Wikipedia documents for a particular date as follows.

```
$ wget http://dumps.wikimedia.org/enwiki/20140903/
$ bzip2 -d enwiki-20140903-pages-articles-multistream.xml.bz2
```

We index a large document corpus with Solr's data import handler. First, the handler specifies the XML fields in the corpus and maps fields that are specific to Wikipedia to those used by Solr.

```
<dataConfig>
   <dataSource type="FileDataSource" encoding="UTF-8" />
   <document>
   <entity name="page"
           processor="XPathEntityProcessor"
           stream="true"
           forEach="/mediawiki/page/"
           url="enwiki-20140903-pages-articles-multistream.xml"
```

```
        >
    <field column="id          xpath="/mediawiki/page/id" />
    <field column="title"      xpath="/mediawiki/page/title" />
    <field column="revision"   xpath="/mediawiki/page/revision/id" />
<field column="user"      xpath="/mediawiki/page/contributor/username" />
    <field column="userId"     xpath="/mediawiki/page/contributor/id" />
    <field column="text"       xpath="/mediawiki/page/text" />
    <field column="timestamp" xpath="/mediawiki/page/timestamp" />
    </entity>
    </document>
</dataConfig>
```

Second, after defining the configuration file, we include the library with the data import handler into the Solr configuration file `solrconfig.xml`, and we register `data-config.xml`.

```
<lib dir=.''./../../dist/" regex="solr-dataimporthandler-.*\.jar" />

[....]

<requestHandler
   name="/dataimport"
   class="org.apache.solr.handler.dataimport.DataImportHandler">
   <lst name="defaults">
      <str name="config">data-config.xml</str>
   </lst>
</requestHandler>
```

Finally, we perform the data import and create an index for the Wikipedia dataset with the following command within the guest machine. Indexing requires three to four hours for the Wikipedia document corpus, which is nearly 50 GB in size. The index is saved in the directory for `collection1`. Once indexing is complete, we use cURL commands to issue queries against a real-world document corpus and simulate its behavior as described earlier.

```
$ curl "http://localhost:8983/solr/collection1/
     dataimport?command=full-import"
```

5.8 ADDITIONAL CONSIDERATIONS

In this chapter, we have described infrastructure for full-system simulation and its application to datacenter workloads. We have emphasized, in particular, the support for virtual machine snapshots that allows precisely targeted checkpoints and simulations. Checkpoints and careful transitions improve the tractability of cycle-level simulation for datacenter workloads by de-limiting regions of interest. Checkpoints and the seamless flow between emulation and simulation modes enhance analysis speed by incurring simulation costs only for precisely defined, domain-specific application tasks (e.g., a search query). Because these tasks are short and complete in tens of milliseconds, simulation costs per task are tractable.

This approach to simulation—short and precisely defined tasks—differ from the simulation challenges and solutions found by researchers in the 2000s. Conventional processor benchmarks were large, monolithic applications. For example, simulating a SPEC CPU benchmark would require billions of instructions and days of computation. To address this challenge, researchers proposed a series of methods to accelerate the simulation of long-running workloads. Methods—such as SimPoints [130, 145, 146], TurboSmarts [161], and others—analyze application phase behavior to identify representative instruction segments for simulation. And predictive techniques—such as analytical models [45, 78], regression models [89, 90], neural networks [72]—reduced the number of simulations required for design analysis and optimization.

Datacenter tasks are short and getting shorter [123], reducing the need to identify short, representative instruction segments for simulation. However, datacenters do pose new challenges in identifying representative tasks for simulation. For example, search engines provide a rich set of operators for query construction. Determining the mix of queries that accurately represent workload behavior becomes the challenge. Indeed, the effectiveness of simulating individual datacenter tasks must be predicated on running many of these tasks according to realistic frequency distributions to produce realistic performance distributions.

Finally, we note that much work remains in precisely isolating datacenter computation for simulators. Datacenter workloads increasingly perform computation within sophisticated software frameworks and run-time systems. When the run-time system and tasks run in the same simulated machine, the simulator inevitably captures activity from both. These effects arise in, for example, Spark computation when the task scheduler and tasks share the same machine [171]. Separating the run-time manager and the workers is possible (e.g., with multi-node simulation) but doing so introduces new challenges in managing communication.

Thus, advances in full-system simulation have produced speed and tractability for computer architects who wish to simulate datacenter workloads. Indeed, the difficulty is less often the simulator and more often the domain-specific expertise required to deploy an application and exercise that application with realistic data. Nonetheless, computer architects have sophisticated infrastructure to understand hardware behavior within a datacenter server.

CHAPTER 6
System Simulation

Server-level simulation provides insight into processor and memory system behavior for individual tasks and queries in a datacenter application. Hardware design focuses on narrowly defined figures of merit, such as instruction throughput, average memory access time. In contrast, datacenter management focuses on broadly defined measures of behavior, such as response time distributions. Closing the gap between these design and management objectives means linking results from server-level simulation and inputs into datacenter-level analysis.

At datacenter scale, individual task performance affects queueing dynamics and application service quality. Architects can reason about queueing delays with rules of thumb, such as Little's Law, or abstract models for specific queueing scenarios, such as M/M/1 models. Beyond these simple models, architects can employ discrete event simulators that track tasks as they arrive according to arbitrarily general distributions on inter-arrival times and as they are serviced according to server-level models of task performance.

Given task arrival and service times, architects can employ discrete event simulators to reason about sophisticated queuing dynamics, account for task parallelism at scale, and assess response time distributions. The response time distribution is a cornerstone in datacenter performance analysis. The cumulative distribution function (CDF), which measures the percentage of tasks that complete within a particular target for response time, is instrumental when reasoning about task percentiles or latency tails.

Finally, a more integrated approach is possible. With full-system simulation, an architect could launch multiple instances of a simulator, assign a network address to each instance, and connect the simulators as servers would be connected in a datacenter. This strategy resembles the approach taken for many-core chip-multiprocessors [114], another setting in which architects require scalability when studying communication between coordinated computational tasks.

6.1 ANALYTICAL QUEUEING MODELS

Queueing models provide a framework for reasoning about task dynamics in terms of arrival rates, service times, and end-to-end response times. Queueing models are particularly attractive for first-order, back-of-the-envelope calculations because they can concisely describe task performance with closed-form expressions. But models need to be used with care because they make specific assumptions about system activity.

We present the basic ideas behind queueing models and illustrate how these models have been used to understand datacenter dynamics. A detailed treatment of queueing models is beyond the scope of this synthesis lecture. For further detail, please see Trivedi's treatment of probability, statistics, and queueing [156] or Harchol-Balter's treatment of queueing theory [64]. Both of these books are written with several examples and applications to computer systems.

M/M/1 Queueing Model. The simplest M/M/1 queue provides an easily understood abstraction for the system and is often the starting point for understanding datacenter dynamics. M/M/1 queues model a system in which a server performs computation for tasks as they arrive. Task arrivals form a Poisson process with rate λ. In other words, task inter-arrival times follow an exponential distribution with mean $1/\lambda$. The server performs computation with service times that follow an exponential distribution with mean μ. The system enforces first-in-first-out (FIFO) queueing discipline.

Such a simple model works well when modeling a small system with little concurrency, such as a single server or processor. The model also formalizes a number of system properties. First, a system is unstable when $\lambda > \mu$ since the number of queued tasks grows without bound when the arrival rate exceeds the service rate. In contrast, the system is stable when $\lambda < \mu$ and $\rho = \lambda/\mu$ is a measure of server utilization.

Furthermore, we can employ closed-form expressions for key measures of system performance. If N is a random variable for the number of tasks in the system, its mean and variance are given by the following

$$\begin{aligned} E[N] &= \frac{\rho}{1-\rho} \\ Var[N] &= \frac{\rho}{(1-\rho)^2}. \end{aligned}$$

Given these summary statistics for the number of tasks in the system, we can estimate response time R. We can use Little's Law, which states that expected number of tasks in the system is the product of task arrival rate and the expected time each task spends in the system $E[R]$. Given this relationship, the expected response time can be expressed in terms of the arrival and service rates.

$$\begin{aligned} E[N] &= \lambda E[R] \\ E[R] &= E[N]\lambda^{-1} = \frac{1}{\mu-\lambda}. \end{aligned}$$

Little's Law and the expression for a task's mean response time provide standard rules of thumb. Yet datacenter managers must often look beyond the mean and consider the full response time distribution. Specifically, datacenters often define service quality in terms of response time percentiles. For example, if the datacenter wishes to ensure 99% of tasks should complete before 50 ms, the 99-th percentile response time should be less than 50 ms. For an M/M/1 queue, the p-th percentile response time can be expressed in closed-form.

$$R_p = -\frac{\ln(1-p/100)}{\mu-\lambda}.$$

Beyond understanding task dynamics in the system, we may wish to understand task dynamics in the queue. We can estimate a task's waiting time W in the queue, which is the difference between response time R and service time S. Once we have expected waiting time, we can use Little's Law to obtain expected queue occupancy Q. Note that the difference between system occupancy and queue occupancy must be server occupancy (i.e., utilization ρ).

$$\begin{aligned} E[W] &= E[R] - E[S] = \frac{1}{\mu - \lambda} - \frac{1}{\mu} \\ E[Q] &= \lambda E[W] \\ E[N] - E[Q] &= \rho. \end{aligned}$$

M/M/k Queueing Model. Datacenters are comprised of many servers, but the M/M/1 queueing model considers a single server. The M/M/1 model extends naturally to a system with multiple (k) servers with a corresponding number of queues as long as each server draws tasks from its own queue (i.e., no work stealing). The only question is how to distribute tasks to the queues and servers. If tasks are distributed uniformly at random to homogeneous servers, M/M/1 models provide insight into the queueing dynamics at each server.

Alternatively, the M/M/k queueing model considers a system with multiple (k) servers that draw tasks from a shared queue. Tasks arrive according to a Poisson process with arrival rate λ and each server computes for tasks with service rate μ. Unfortunately, this system is more difficult to analyze in closed form. The following illustrates model complexity for the expected number of tasks in the system N where $\rho = \lambda/(k\mu)$.

$$\begin{aligned} \pi_0 &= \left[\sum_{i=0}^{k-1} \frac{(k\rho)^i}{i!} + \frac{(k\rho)^k}{k!} \frac{1}{1-\rho} \right]^{-1} \\ E[N] &= k\rho + \rho \frac{(k\rho)^k}{k!} \frac{\pi_0}{(1-\rho)^2}. \end{aligned}$$

Given the complexity of the M/M/k model, a natural question arises. In a datacenter with k servers, should we use multiple M/M/1 queues or a single M/M/k queue? Trivedi addresses this question in his book [156] with an example that models an operating system that manages two processors. He compares the expected response time from two organizations: (i) two processors with two queues and (ii) two processors with a shared queue.

First, two processors with two queues correspond to two M/M/1 queueing systems, each with half of the task load ($\lambda/2$). Thus, we compute the expected response time using a standard application of the M/M/1 model.

$$E[R_{M/M/1}] = \frac{1}{\mu - \lambda/2} = \frac{2}{2\mu - \lambda}.$$

Second, two processors with a shared queue corresponds to an M/M/2 queueing system. We calculate the expected response time by calculating the expected number of tasks in the system

($\rho = \lambda/(2\mu)$) and applying Little's Law.

$$\begin{aligned} \pi_0 &= \left[1 + 2\rho + \frac{(2\rho)^2}{2!}\frac{1}{1-\rho}\right]^{-1} = \frac{1-\rho}{1+\rho} \\ E[N] &= 2\rho + \frac{\rho(2\rho)^2}{2!}\frac{\pi_0}{(1-\rho)^2} = \frac{2\rho}{1-\rho^2} \\ E[R_{M/M/2}] &= E[N]\lambda^{-1} = \frac{4\mu}{4\mu^2-\lambda^2}. \end{aligned}$$

When we compare expected response times, we find that those for multiple queues are higher than those for a single pooled queue.

$$E[R_{M/M/1}] = \frac{2}{2\mu-\lambda} = \frac{4\mu+2\lambda}{4\mu^2-\lambda^2} > E[R_{M/M/2}].$$

Although the queueing theory states that a single queue of pooled tasks performs better than multiple queues with distributed tasks, datacenter task schedulers are likely to implement the former rather than the latter. At warehouse-scale, datacenter schedulers may suffer intolerable performance penalties when implementing a globally shared queue. A combination of hierarchical scheduling, in which multiple frameworks implement their own queueing disciplines other than FIFO (e.g., Mesos [67]) and multi-scale scheduling in which a framework schedules jobs, operators, and tasks at progressively finer granularities (e.g., Spark [171]), means that datacenters may be too complex to model with M/M/1 queues and instead should be modeled as networks of queues, which are not amenable to closed-form analysis.

Despite their limitations M/M/1 and M/M/k queues are relevant to datacenter researchers. They provide quick rules of thumb and first order estimates on system performance. Moreover, they complement cycle-level server simulation in the previous chapter. Processor and memory simulations provide estimates of task service time, which provide inputs to queueing models. Combined with a model for task arrivals, usually some diurnal activity pattern, queueing models can translate task performance into system performance.

6.2 DISCRETE EVENT SIMULATION

The M/M/1 and M/M/k models make restrictive assumptions about the statistical properties of task arrivals and service times. In general, the statistical distributions for inter-arrival times and service times may deviate from the exponential distribution. Perhaps inter-arrival times exhibit greater variance. Or perhaps service times are better characterized by multi-modal distributions with modes that correspond to multiple types of heterogeneous servers. In these settings, the clean assumptions required for M/M/1 models are unrealistic.

For system dynamics that do not conform to assumptions on statistical distributions, we require the generality of G/G/k queueing systems. The first G denotes a general distribution for inter-arrival times and the second G denotes a general distribution for service times. The k denotes the number of servers. Generalized distributions preclude closed-form expressions for performance. Instead, performance analysts must rely on approximations with variable fidelity or turn to discrete event simulation.

Discrete event simulation tracks the flow of tasks through a system. The simulator requires two inputs—arrival times and service times. Task arrivals could be drawn from time-stamped traces, which provide an empirical distribution on inter-arrival times. Or they could be drawn from a standard exponential distribution as in the M/M/1 queueing system. Service times are drawn from physical measurements for real workloads running on real hardware. Or they can be drawn from cycle-level simulations for new processor and memory designs. Because arrival and service times exhibit variance, a performance analyst should construct and sample from an empirical probability distribution when providing inputs to the simulator. With arbitrary, empirical distributions for both arrival and service times, discrete event simulation provides analytical flexibility.

In exchange for flexibility, discrete event simulation requires additional computation. The simulator tracks individual tasks as they arrive, queue, compute, and complete. The simulator places arriving tasks and their associated arrival timestamps into the priority queue. When a server becomes available, the task at the head of the queue begins computation. The simulator places a completion event and its associated timestamp, which is computed based on the empirical distribution on service times, into the priority queue. The priority queue orders task events by timestamps, which enforces FIFO queueing discipline. The simulator collects and updates timestamps for each task's key events—arrival, dequeue, completion. In aggregate, timestamps across all tasks provide the requisite measures of datacenter performance, including summary statistics and percentiles for waiting time, service time, and end-to-end response time.

Meisner et al. implement a discrete event queueing simulator, BigHouse, which provides much of the functionality we describe in this section [50, 112]. They describe an infrastructure that profiles and logs a deployed system, uses these profiles to create a workload model that defines task arrival and service distributions, and launches a stochastic queueing simulator. Meisner et al. devote particular attention to sampling the statistics reported by the simulator. Observations of waiting time, service time, and response time for consecutive tasks are correlated. Intuitively, one task's long service time affects another task's waiting time. To avoid this bias, Meisner et al. sample observations that are separated in time to produce independent and identically distributed (i.i.d.) data from the simulator.

6.3 PARALLEL SYSTEM SIMULATION

Thus far, we have separated server- and datacenter-level simulators with task service times to provide a clean link between the two. Modularity improves tractability. For example, we could simulate multiple server designs with varied processors and memories. We could then simulate a heterogeneous datacenter that combines the varied server designs into a single system. The discrete event simulator would provide insight into the queueing behaviors that arise from heterogeneous servers, a study that would be possible but rather daunting with analytical models.

Limitations of Decoupled Simulation. Although decoupled simulation provides tractability, it has a number of limitations. First, decoupled simulation focuses on computation and neglects other aspects of system behavior. Server-level simulations estimate service time by modeling application performance based on processor and memory parameters. We emulate network and storage, which produces functionally correct results but neglects the performance impact of communication and data movement. Yet these effects are likely modest. In practice, well tuned datacenter applications exhibit abundant task parallelism and should rarely communicate with storage. Web search sizes document indexes to fit within main memory. Relative to Hadoop MapReduce, Spark improves the performance of iterative machine learning codes by one or two orders of magnitude by caching data in main memory across iterations.

Moreover, decoupled simulation focuses on task parallelism and neglects interactions and dependences between tasks. Certainly, task schedulers construct data dependence graphs and enforce those dependences when issuing tasks to workers. But decoupled simulation does not account for phase changes in application behavior. For example, Hadoop MapReduce performs computation for Map tasks, communicates intermediate results with a data shuffle through the distributed file system, and performs computation for Reduce tasks. Our simulation approach captures worker performance and queueing dynamics of the Map and Reduce phases, but does not account for shuffle performance. These effects are likely modest, especially as more and more data is cached in main memory, but may produce performance outliers and stragglers in real systems that would be missed in our simulated system.

Distributed Simulation. One approach that addresses these limitations is parallel system simulation, which launches and coordinates the simultaneous simulation of multiple servers. For example, we could launch two instances of MARSSx86/QEMU, take note of their network addresses, and register both machines as part of a datacenter application's deployment. The simulated servers would typically perform independent computation on tasks, but may communicate with each during data shuffles. To accurately assess communication's effect on server performance, however, MARSSx86 and QEMU would need new timing models for the network interface, in addition to those for the processor and memory system. These models have been developed for other simulators, such as M5 [22], to support research in TCP/IP networking.

Parallel system simulation has also been demonstrated for many-core research. The Graphite simulator is a distributed parallel multi-core simulator designed to evaluate tens or hundreds of cores [114]. Graphite automatically distributes threads and simulated cores of the target architecture across multiple host machines, while maintaining the illusion of a single, shared address space. To ensure tractable performance at such scale, Graphite simplifies the timing models (i.e., the simulator is not cycle-accurate) and relies on analytical approximations.

Sniper further accelerates Graphite with interval simulation, a technique that raises the level of abstraction [28, 53]. Rather than track each instruction's progress through pipeline stages, it tracks contiguous intervals of instruction flow separated by disruptive events such as branch misprediction, cache misses, and TLB misses. Fast analytical models estimate interval performance, and detailed simulators model the effect of disruptive events. The coupled model and simulator provides accuracy and speed.

Functionally, MARSSx86/QEMU could provide cycle-level performance analysis across distributed machines, but simulator performance may suffer. Efficient and tightly coupled simulation of datacenter servers remains a challenge in research methodology.

CHAPTER 7

Conclusions

Research in computer architecture has evolved significantly over the past two decades. In the 1990s, research focused on microarchitectural mechanisms to extract instruction-level parallelism for general-purpose computing. In comparison, today's research is notable for its hardware breadth and software depth. Architects look beyond the processor core organization, which determines only a fraction of system capability, in the pursuit of complementary advances for memory organization and system composition. Moreover, architects target specific software frameworks and big data applications in the pursuit of larger gains in performance and power efficiency via specialization. Taken together, a broader view of hardware and a deeper view of software provides many perspectives on datacenter and server design. One goal of this book has been to survey representative software frameworks and the hardware platforms designed for them.

Today's research in computer architecture is also notable for its contributions to datacenter management. With her deep insight into hardware behavior, a computer architect provides unique perspectives on scheduling tasks, allocating hardware, enforcing allocations, and diagnosing performance pathologies. Moreover, design and management are inextricably linked. An architect cannot simply design a processor and wait for system adoption. Rather, she should anticipate management challenges during design, which might mean moderating design extrema to reduce the risk of poor management and performance. A second goal of this book has been to highlight datacenter management from a hardware architect's perspective.

Notwithstanding recent advances, computer architects encounter numerous methodological hurdles when studying datacenters. First, evaluating prospective architectures and technologies for datacenter applications requires expensive, full-system simulation. Researchers must integrate detailed timing models for hardware with a complete system stack for software. The software stack includes an emulator, an operating system, a run-time system (e.g., Hadoop or Spark), a sophisticated application benchmark, and a realistic dataset. Second, because only a few, large entities possess warehouse-scale datacenters, researchers should supplement server-level simulations with datacenter-level ones. A third goal of this book has been to illustrate experimental infrastructure that have proven effective for datacenter-oriented architecture research.

Future Trends in Software. Datacenters will deploy increasingly sophisticated system software for distributed computing. These software frameworks will look beyond map and reduce functions to a broad range of transformations and functions on data sets. More capable computational primitives permit expressive languages and powerful libraries with which application developers can quickly deploy distributed and parallel jobs. Many of these programming frameworks may be domain-specific yet balance programmability and efficiency.

Beyond new capabilities, datacenter software will benefit from new performance optimizations. The past decade produced new programming models that democratized distributed computing and derived performance from scale. In contrast, the next decade will produce new performance optimizations that qualitatively improve latency and throughput at a given scale through shorter and fine-grained tasks, sophisticated run-times for tasks scheduling and resource allocation, as well as intelligent data management and communication. An early example of such optimizations is caching data in local memory rather than retrieving it from a distributed file system. Performance optimizations will shift datacenter activity from network and storage to processors and memory, parts of the hardware platform that computer architects know well.

Future Trends in Hardware. Datacenters will deploy increasingly heterogeneous hardware platforms. System efficiency will benefit when computation uses a mix of small and big processor cores. It will further benefit when computation exploits accelerators, whether they be graphics processors, field-programmable gate arrays, or application-specific integrated circuits. System efficiency and reliability will benefit when data resides in hybrid memory systems with a mix of high-bandwidth and low-power interfaces and a mix of classical technologies and emerging non-volatile ones. Finally, system responsiveness will benefit when network protocols co-design system software and server hardware. In this future, organizing components into a coherent and balanced system will be as important as the design of individual components (if not more so).

Datacenters may demand more than performance and power efficiency from its hardware. Design objectives may expand to include emerging priorities such as performance reliability and security. First, managing the variance in task performance can be viewed from the perspective of reliability. As datacenters seek to mitigate and shorten tails in their tasks' latency distributions, their objectives may resemble those in real-time systems, which explicitly optimize designs to reduce worst-case execution time. Hardware support may prove instrumental in guaranteeing response time and bounding worst case latency in datacenters.

Second, encouraging user participation in cloud computing will require new incentives and security guarantees. A user who computes in a shared system will seek game-theoretic assurances. Without such assurances, strategic users might manipulate system management policies to benefit themselves at the expense of others. Such behavior has long been reported in shared high-performance computing clusters, and new hardware management frameworks will be needed to avoid similar outcomes in cloud computing. Users will also seek security assurances as they migrate to the cloud and relinquish physical control of their applications and data. Hardware support will be necessary to guarantee data privacy and integrity.

Concluding Remarks. In summary, computer architects who are interested in datacenters will find many research avenues to follow. Large computer systems have existed for decades, but modern datacenters (i.e., those that use commodity hardware and perform computation with massive task parallelism) have emerged relatively recently. During this same period, the role of computer architecture has become increasingly important. The end of Dennard scaling and the slowing of Moore's Law means that datacenter operators can no longer rely on technology for faster, more efficient components. The emergence of new technologies means that datacenter operators will need to adapt existing system organizations and explore new design spaces. Although much has been accomplished, much more research is needed to navigate new challenges and exploit new opportunities at the intersection of computer architecture and datacenters.

Bibliography

[1] T. Aamodt et al. GPGPU-Sim. http://www.gpgpu-sim.org/. Accessed: 2015-06-09. 56, 60

[2] D. Andersen, J. Franklin, M. Kaminsky, A. Phanishayee, L. Tan, and V. Vasudevan. FAWN: A fast array of wimpy nodes. In *Proc. Symposium on Operating Systems Principles (SOSP)*, 2009. DOI: 10.1145/1629575.1629577. 14, 37, 38

[3] D. Andersen, J. Franklin, M. Kaminsky, A. Phanishayee, L. Tan, and V. Vasudevan. FAWN: A fast array of wimpy nodes. *Communications of the ACM (CACM)*, 2011. DOI: 10.1145/1965724.1965747. 14

[4] Anonymous. Space invaders. *The Economist*, 2012. 56

[5] Apache. Giraph. http://giraph.apache.org/. Accessed: 2015-02-03. 20

[6] Apache. GraphX. https://spark.apache.org/graphx/. Accessed: 2015-02-03. 20

[7] Apache. Hbase. http://hbase.apache.org/. Accessed: 2015-02-03. 18

[8] Apache. Hive. http://hive.apache.org/. Accessed: 2015-02-03. 18

[9] Apache. Lucene. http://lucene.apache.org/. Accessed: 2015-02-03. 25, 73

[10] Apache. Mahout. http://mahout.apache.org/. Accessed: 2015-02-03. 18

[11] Apache. Nutch. http://nutch.apache.org/. Accessed: 2015-02-03. 25, 73, 76

[12] Apache. Pig. http://pig.apache.org/. Accessed: 2015-02-03. 18

[13] Apache. Spark. http://spark.apache.org/. Accessed: 2015-02-03. 18

[14] B. Atikoglu, Y. Xu, E. Frachtenberg, S. Jiang, and M. Paleczny. Workload analysis of a large-scale key-value store. In *Proc. SIGMETRICS Performance Evaluation Review*, 2012. DOI: 10.1145/2318857.2254766. 5, 13, 15

[15] W. Baek and T. Chilimbi. Green: A framework for supporting energy-conscious programming using controlled approximations. In *Proc. Conference on Programming Language Design and Implementation (PLDI)*, 2010. DOI: 10.1145/1806596.1806620. 31

[16] L. Barroso, J. Clidaras, and U. Hoelzle. The datacenter as a computer: An introduction to the design of warehouse-scale machines. *Synthesis Lectures on Computer Architecture*, 2013. DOI: 10.2200/s00193ed1v01y200905cac006. xi, 24, 31

[17] L. Barroso, J. Dean, and U. Hoelzle. Web search for a planet: The Google cluster architecture. *Micro*, 2003. DOI: 10.1109/mm.2003.1196112. 23

[18] L. Barroso, K. Gharachorloo, R. McNamara, A. Nowatzyk, S. Qadeer, B. Sano, S. Smith, R. Stets, and B. Verghese. Piranha: A scalable architecture based on single-chip multiprocessing. In *Proc. International Symposium on Computer Architecture (ISCA)*, 2000. DOI: 10.1109/isca.2000.854398. 24

[19] S. Beamer, K. Asanovic, and D. Patterson. Direction-optimizing breadth-first search. In *Proc. International Conference on High Performance Computing, Networking, Storage and Analysis (SC)*, 2012. DOI: 10.1109/sc.2012.50. 21

[20] C. Bienia. Benchmarking modern multiprocessors. In Ph.D. Thesis, *Princeton University*, 2011. 22

[21] N. Binkert, B. Beckmann, G. Black, et al. The gem5 simulator. *SIGARCH Computer Architecture News*, 2011. DOI: 10.1145/2024716.2024718. 56, 57, 60

[22] N. Binkert, R. Dreslinski, L. Hsu, K. Lim, A. Saidi, and S. Reinhardt. The M5 simulator: Modeling networked systems. *IEEE Micro*, 2006. DOI: 10.1109/mm.2006.82. 84

[23] E. Bragg, M. Guevara, and B. Lee. Understanding query complexity and its implications for energy-efficient web search. In *Proc. International Symposium on Low Power Electronics and Design (ISLPED)*, 2013. DOI: 10.1109/islped.2013.6629330. 9, 25, 26, 28, 73

[24] S. Brin and L. Page. The anatomy of a large-scale hypertextual web search engine. *Computer Networks*, 1998. DOI: 10.1016/j.comnet.2012.10.007. 6

[25] D. Brooks, V. Tiwari, and M. Martonosi. Wattch: A framework for architectural-level power analysis and optimizations. In *Proc. International Symposium on Computer Architecture (ISCA)*, 2000. DOI: 10.1145/339647.339657. 56

[26] A. Buluc and J. Gilbert. The Combinatorial BLAS: Design, implementation, and applications. In *Proc. International Symposium on High Performance Computer Architecture (HPCA)*, 2011. DOI: 10.1177/1094342011403516. 20

[27] D. Burger and T. Austin. The SimpleScalar tool set, version 2.0. *SIGARCH Computer Architecture News*, 25(3), 1997. DOI: 10.1145/268806.268810. 55

[28] T. Carlson, W. Heirman, and L. Eeckhout. Sniper: Exploring the level of abstraction for scalable and accurate parallel multi-core simulation. In *Proc. International Conference for High Performance Computing, Networking, Storage and Analysis (C)*, 2011. DOI: 10.1145/2063384.2063454. 84

[29] F. Chang, J. Dean, S. Ghemawat, W. Hsieh, D. Wallach, M. Burrows, T. Chandra, A. Fikes, and R. Gruber. Bigtable: A distributed storage system for structured data. In *Proc. Symposium on Operating Systems Design and Implementation (OSDI)*, 2006. DOI: 10.1145/1365815.1365816. 18

[30] J. Chase, D. Anderson, P. Thakar, A. Vahdat, and R. Doyle. Managing energy and server resources in hosting centers. In *Proc. Symposium on Operating System Principles (SOSP)*, 2001. DOI: 10.1145/502034.502045. 47, 48

[31] Y. Chen, T. Luo, S. Liu, S. Zhang, L. He, J. Wang, L. Li, T. Chen, Z. Xu, N. Sun, and O. Temam. DaDianNao: A machine-learning supercomputer. In *Proc. International Symposium on Microarchitecture (MICRO)*, 2014. DOI: 10.1109/micro.2014.58. 22

[32] J. Condit, E. Nightingale, C. Frost, E. Ipek, B. Lee, D. Burger, and D. Coetzee. Better I/O through byte-addressable, persistent memory. In *Proc. International Symposium on Operating Systems Principles (SOSP)*, 2009. DOI: 10.1145/1629575.1629589. 37

[33] J. Davis, J. Laudon, and K. Olukotun. Maximizing CMP throughput with mediocre cores. In *Proc. International Conference on Parallel Architectures and Compilation Techniques (PACT)*, 2005. DOI: 10.1109/pact.2005.42. 24

[34] J. Dean and L. Barroso. The tail at scale. *Communications of the ACM*, 2013. DOI: 10.1145/2408776.2408794. 49

[35] J. Dean and S. Ghemawat. MapReduce: Simplified data processing on large clusters. In *Proc. Symposium on Operating Systems Design and Implementation (OSDI)*, 2004. DOI: 10.1145/1327452.1327492. 15, 16, 18

[36] G. DeCandia, D. Hastorun, M. Jampani, G. Kakulapati, A. Lakshman, A. Pilchin, et al. Dynamo: Amazon's highly available key-value store. In *Proc. Symposium on Operating Systems Principles (SOSP)*, 2007. DOI: 10.1145/1294261.1294281. 13

[37] C. Delimitrou and C. Kozyrakis. Paragon: QoS-aware scheduling for heterogeneous datacenters. In *Proc. International Conference on Architectural Support for Programming Languages and Operating Systems (ASPLOS)*, 2013. DOI: 10.1145/2451116.2451125. 30, 45, 46

[38] C. Delimitrou and C. Kozyrakis. Quasar: Resource-efficient and QoS-aware cluster management. In *Proc. International Conference on Architectural Support for Programming Languages and Operating Systems (ASPLOS)*, 2014. DOI: 10.1145/2541940.2541941. 45, 46

[39] Q. Deng, D. Meisner, L. Ramos, T. Wenisch, and R. Bianchini. MemScale: Active low-power modes for main memory. In *Proc. International Conference on Architectural Support for Programming Languages and Operating Systems (ASPLOS)*, 2011. DOI: 10.1145/1950365.1950392. 33

[40] B. Diniz, D. Guedes, and R. Bianchini. Limiting the power consumption of main memory. In *Proc. International Symposium on Computer Architecture*, 2007. DOI: 10.1145/1250662.1250699. 34

[41] A. Diwan. Keynote: Life lessons and datacenter performance analysis. In *Proc. International Symposium on Performance Analysis of Systems and Software (ISPASS)*, 2014. DOI: 10.1109/ispass.2014.6844478. 9

[42] L. Eeckhout. Computer architecture performance evaluation methods. *Synthesis Lectures on Computer Architecture*, 2010. DOI: 10.2200/s00273ed1v01y201006cac010. xi

[43] H. Esmaeilzadeh, A. Sampson, L. Ceze, and D. Burger. Architecture support for disciplined approximate programming. In *Proc. International Conference on Architectural Support for Programming Languages and Operating Systems (ASPLOS)*, 2012. DOI: 10.1145/2150976.2151008. 31

[44] S. Eyerman and L. Eeckhout. System-level performance metrics for multiprogram workloads. *IEEE Micro*, 2008. DOI: 10.1109/mm.2008.44. 48

[45] S. Eyerman, L. Eeckhout, T. Karkhanis, and J. Smith. A performance counter architecture for computing accurate CPI components. In *Proc. International Conference on Architectural Support for Programming Languages and Operating Systems (ASPLOS)*, 2006. DOI: 10.1145/1168857.1168880. 78

[46] X. Fan, C. Ellis, and A. Lebeck. Memory controller policies for DRAM power management. In *Proc. International Symposium on Low Power Electronics and Design (ISLPED)*, 2001. DOI: 10.1145/383082.383118. 34

[47] M. Ferdman, A. Adileh, O. Kocberber, M. Alisafaee S. Volos, D. Jevdjic, C. Kaynak, A. Ailamaki A.D. Popescu, and B. Falsafi. Quantifying the mismatch between emerging scale-out applications and modern processors. *ACM Transactions on Computer Systems*, 2012. DOI: 10.1145/2382553.2382557. 6, 57

[48] M. Ferdman, A. Adileh, O. Kocberber, M. Alisafaee S. Volos, D. Jevdjic, C. Kaynak, A. Ailamaki A.D. Popescu, and B. Falsafi. A case for specialized processors for scale-out workloads. *IEEE Micro*, 2014. DOI: 10.1109/mm.2014.41. 6

[49] M. Ferdman, A. Adileh, O. Kocberber, S. Volos, M. Alisafaee, D. Jevdjic, C. Kaynak, A. Popescu, A. Ailamaki, and B. Falsafi. Clearing the clouds: A study of emerging

scale-out workloads on modern hardware. In *Proc. International Conference on Architectural Support for Programming Languages and Operating Systems (ASPLOS)*, 2012. DOI: 10.1145/2150976.2150982. 6, 32

[50] G. Fishman. *Discrete-event Simulation*. Springer-Verlag, 2001. DOI: 10.1007/978-1-4757-3552-9. 83

[51] R. Gabor, S. Weiss, and A. Mendelson. Fairness and throughput in switch on event multi-threading. In *Proc. International Symposium on Microarchitecture*, 2006. DOI: 10.1109/micro.2006.25. 48

[52] R. Gabor, S. Weiss, and A. Mendelson. Fairness enforcement in switch on event multithreading. *ACM Transactions on Architecture and Code Optimization*, 2007. DOI: 10.1145/1275937.1275939. 48

[53] D. Genbrugge, S. Eyerman, and L. Eeckhout. Interval simulation: Raising the level of abstraction in architectural simulation. In *Proc. International Symposium on High Performance Computer Architecture (HPCA)*, 2010. DOI: 10.1109/hpca.2010.5416636. 84

[54] A. Ghodsi, M. Zahari, B. Hindman, A. Konwinski, S. Shenker, and I. Stoica. Dominant resource fairness: Fair allocation of multiple resource types. In *Proc. Conference on Networked Systems Design and Implementation*, 2011. 50

[55] Google. Inside search. `https://www.google.com/insidesearch/`. Accessed: 2015-02-3. 6, 7

[56] B. Grot, D. Hardy, P. Lotfi-Kamran, C. Nicopoulos B. Falsafi, and Y. Sazeides. Optimizing data-center TCO with scale-out processors. *IEEE Micro*, 2012. DOI: 10.1109/mm.2012.71. 6

[57] C. Guestrin. Slides: Large-scale machine learning and graphs. `http://www.slideshare.net/AmazonWebServices/graphlab-largescale-machine-learning-on-graphs-bdt204-aws-reinvent-2013`, 2013. 69

[58] M. Guevara, B. Lubin, and B.C. Lee. Navigating heterogeneous processors with market mechanisms. In *Proc. International Symposium on High-Performance Computer Architecture (HPCA)*, 2013. DOI: 10.1109/hpca.2013.6522310. 29, 47, 48

[59] M. Guevara, B. Lubin, and B.C. Lee. Strategies for anticipating risk in heterogeneous system design. In *Proc. International Symposium on High-Performance Computer Architecture (HPCA)*, 2014. DOI: 10.1109/hpca.2014.6835926. 30

[60] A. Gutierrez, M. Cieslak, B. Giridhar, R. Dreslinski, L. Ceze, and T. Mudge. Integrated 3D-stacked server designs for increasing physical density of key-value stores. In *Proc. International Conference on Architectural Support for Programming Languages and Operating Systems (ASPLOS)*, 2014. DOI: 10.1145/2541940.2541951. 38

[61] T. Ham, B. Chelepalli, and N. Xue nad B. Lee. Disintegrated control for power-efficient and heterogeneous memory systems. In *Proc. International Symposium on High-Performance Computer Architecture (HPCA)*, 2013. DOI: 10.1109/hpca.2013.6522338. 36, 39

[62] R. Hameed, W. Qadeer, M. Wachs, O. Azizi, A. Solomatnikov, B. Lee, S. Richardson, C. Kozyrakis, and M. Horowitz. Understanding sources of inefficiency in general-purpose chips. In *Proc. International Symposium on Computer Architecture (ISCA)*, 2010. DOI: 10.1145/1815961.1815968. 31

[63] R. Hameed, W. Qadeer, M. Wachs, O. Azizi, A. Solomatnikov, B. Lee, S. Richardson, C. Kozyrakis, and M. Horowitz. Understanding sources of inefficiency in general-purpose chips. *Communications of the ACM (CACM)*, 2011. DOI: 10.1145/2001269.2001291. 31

[64] M. Harchol-Balter. *Performance Modeling and Design of Computer Systems: Queueing Theory in Action*. Cambridge, 2013. DOI: 10.1017/cbo9781139226424. 80

[65] J. Hauswald, Y. Kang, M. Laurenzano, Q. Chen, C. Li, R. Dreslinski, T. Mudge, J. Mars, and L. Tang. Djinn: DNN as a service and its implications for future warehouse scale computers. In *Proc. International Symposium on Computer Architecture (ISCA)*, 2015. DOI: 10.1145/2749469.2749472. 22

[66] J. Hauswald, Y. Zhang, M. Laurenzano, C. Li, A. Rovinski, A. Khurana, R. Dreslinski, V. Petrucci, T. Mudge, L. Tang, and J. Mars. Sirius: An open end-to-end voice and vision personal assistant and its implications for future warehouse scale computers. In *Proc. International Conference on Architectural Support for Programming Languages and Operating Systems (ASPLOS)*, 2015. DOI: 10.1145/2694344.2694347. 22

[67] B. Hindman, A. Konwinski, M. Zaharia, et al. Mesos: A platform for fine-grained resource sharing in the data center. In *Proc. Symposium on Networked System Design and Implementation (NSDI)*, 2011. 42, 82

[68] U. Hoelzle. Brawny cores still beat wimpy cores, most of the time. *IEEE Micro*, 2010. 28

[69] R. Hou, T. Jiang, L. Zhang, P. Qi, J. Dong, H. Wang, X. Gu, and S. Zhang. Cost effective data center servers. In *Proc. International Symposium on High-Performance Computer Architecture (HPCA)*, 2013. DOI: 10.1109/hpca.2013.6522317. 37

[70] C. Hsu, Y. Zhang, M. Laurenzano, D. Meisner, T. Wenisch, J. Mars, L. Tang, and R. Dreslinski. Adrenaline: Pinpointing and reining in tail queries with quick voltage boosting. In *Proc. International Symposium on High Performance Computer Architecture (HPCA)*, 2015. DOI: 10.1109/hpca.2015.7056039. 30

[71] Intel Corporation. Improving real-time performance by utilizing cache allocation technology. In Intel white paper, 2015. 32

[72] E. Ipek, S. McKee, B. de Supinski, M. Schulz, and R. Caruana. Efficiently exploring architectural design spaces via predictive modeling. In *Proc. International Conference on Architectural Support for Programming Languages and Operating Systems (ASPLOS)*, 2006. DOI: 10.1145/1168857.1168882. 78

[73] M. Isard, M. Budiu, Y. Yu, A. Birrell, and D. Fetterly. Dryad: Distributed data-parallel programs from sequential building blocks. In *Proc. European Conference of Computer Systems (EuroSys)*, 2007. DOI: 10.1145/1272996.1273005. 18

[74] B. Jansen and A. Spink. How are we searching the World Wide Web? A comparison of nine search engine transaction logs. *Information Processing and Management*, 2006. DOI: 10.1016/j.ipm.2004.10.007. 8, 10

[75] W. Jia, L. Wang, J. Zhan, L. Zhang, and C. Luo. Characterizing data analysis workloads in data centers. In *Proc. International Symposium on Workload Characterization (IISWC)*, 2013. DOI: 10.1109/iiswc.2013.6704671. 6

[76] Z. Jia, J. Zhan, W. Lei, R. Han, S. McKee, Q. Yang, C. Luo, and J. Li. Characterizing and subsetting big data workloads. In *Proc. International Symposium on Workload Characterization (IISWC)*, 2014. DOI: 10.1109/iiswc.2014.6983058. 6

[77] S. Kanev, J. Darago, K. Hazelwood, P. Ranganathan, T. Moseley, G. Wei, and D. Brooks. Profiling a warehouse-scale computer. In *Proc. International Symposium on Computer Architecture (ISCA)*, 2015. DOI: 10.1145/2749469.2750392. 43

[78] T. Karkhanis and J. Smith. Automated design of application specific superscalar processors: An analytical approach. In *Proc. International Symposium on Computer Architecture (ISCA)*, 2007. DOI: 10.1145/1250662.1250712. 78

[79] O. Kocberber, B. Grot, J. Picorel, B. Falsafi, K. Lim, and P. Ranganathan. Meet the walkers: Accelerating index traversals for in-memory databases. In *Proc. International Symposium on Microarchitecture (MICRO)*, 2013. DOI: 10.1145/2540708.2540748. 31

[80] P. Kongetira, K. Aingaran, and K. Olukotun. Niagara: A 32-way multi-threaded Sparc processor. *Micro*, 2005. DOI: 10.1109/mm.2005.35. 24

[81] Y. Koren, R. Bell, and C. Volinsky. Matrix factorization techniques for recommender systems. *IEEE Computer*, 2009. DOI: 10.1109/mc.2009.263. 45

[82] C. Kozyrakis, A. Kansal, S. Sankar, and K. Vaid. Server engineering insights for large-scale online services. *Micro*, 2010. DOI: 10.1109/mm.2010.73. 32

[83] R. Kumar, K. Farkas, N. Jouppi, P. Ranganathan, and D. Tullsen. Single-ISA heterogeneous multi-core architectures: The potential for processor power reduction. In *Proc. International Symposium on Microarchitecture*, 2003. DOI: 10.1109/micro.2003.1253185. 28

[84] R. Kumar, D. Tullsen, and N. Jouppi. Core architecture optimization for heterogeneous chip multiprocessors. In *Proc. International Conference on Parallel Architectures and Compilation Techniques*, 2006. DOI: 10.1145/1152154.1152162.

[85] R. Kumar, D. Tullsen, P. Ranganathan, N. Jouppi, and K. Farkas. Single-ISA heterogeneous multi-core architectures for multithreaded workload performance. In *Proc. International Symposium on Computer Architecture*, 2004. DOI: 10.1109/isca.2004.1310764. 28

[86] H. Kung. Memory requirements for balanced computer architectures. In *Proc. International Symposium on Computer Architecture (ISCA)*, 1986. DOI: 10.1145/17356.17362. 38

[87] A. Kyrola, G. Blelloch, and C. Guestrin. GraphChi: Large-scale graph computation on just a PC. In *Proc. Symposium on Operating System Design and Implementation (OSDI)*, 2012. 19

[88] A. Lebeck, X. Fan, H. Zeng, and C. Ellis. Power aware page allocation. In *Proc. International Conference on Architectural Support for Programming Languages and Operating Systems (ASPLOS)*, 2000. DOI: 10.1145/378993.379007. 34

[89] B. Lee and D. Brooks. Accurate and efficient regression modeling for microarchitectural performance and power prediction. In *Proc. International Conference on Architectural Support for Programming Languages and Operating Systems (ASPLOS)*, 2006. DOI: 10.1145/1168857.1168881. 78

[90] B. Lee and D. Brooks. Illustrative design space studies with microarchitectural regression models. In *Proc. International Symposium on High-Performance Computer Architecture (HPCA)*, 2007. DOI: 10.1109/hpca.2007.346211. 78

[91] B. Lee, E. Ipek, O. Mutlu, and D. Burger. Architecting phase change memory as a scalable DRAM alternative. In *Proc. International Symposium on Computer Architecture (ISCA)*, 2009. DOI: 10.1145/1555754.1555758. 37, 56

[92] B. Lee, E. Ipek, O. Mutlu, and D. Burger. Phase change memory architecture and the quest for scalability. *Communications of the ACM*, 53(7):99–106, July 2010. 37

[93] B. Lee, P. Zhou, J. Yang, Y. Zhang, B. Zhao, E. Ipek, O. Mutlu, and D. Burger. Phase change technology and the future of main memory. *IEEE Micro*, 30(1):131–141, January 2010. 37

[94] J. Leskovec and A. Krevl. SNAP Datasets: Stanford large network dataset collection. `http://snap.stanford.edu/data`. Accessed: 2015-02-3. 21, 38

[95] J. Leskovec and R. Sosic. SNAP: A general purpose network analysis and graph mining library in C++. `http://snap.stanford.edu/`. Accessed: 2015-02-3. 21

[96] S. Li, J. Ahn, R. Strong, et al. McPAT: An integrated power, area, and timing modeling framework for multicore and manycore architectures. In *Proc. International Symposium on Microarchitecture (MICRO)*, 2009. DOI: 10.1145/1669112.1669172. 60

[97] S. Li, H. Lim, V. Lee, J. Ahn, A. Kalia, M. Kaminsky, D. Andersen, S. O, S. Lee, and P. Dubey. Architecting to achieve a billion requests per second throughput on a single key-value store server platform. In *Proc. International Symposium on Computer Architecture (ISCA)*, 2015. DOI: 10.1145/2749469.2750416. 13, 38

[98] H. Lim, D. Han, D. Andersen, and M. Kaminsky. MICA: A holistic approach to fast in-memory key-value storage. In *Proc. Symposium on Networked Systems Design and Implementation (NSDI)*, 2014. 13

[99] K. Lim, J. Chang, T. Mudge, P. Ranganathan, S. Reinhart, and T. Wenisch. Disaggregated memory for expansion and sharing in blade servers. In *Proc. International Symposium on Computer Architecture (ISCA)*, 2009. DOI: 10.1145/1555754.1555789. 37

[100] K. Lim, D. Meisner, A. Saidi, P. Ranganathan, and T. Wenisch. Thin servers with smart pipes: Designing SoC accelerators for memcached. In *Proc. International Symposium on Computer Architecture (ISCA)*, 2013. DOI: 10.1145/2485922.2485926. 15, 40

[101] K. Lim, P. Ranganathan, J. Chang, C. Patel, T. Mudge, and S. Reinhardt. Understanding and designing new server architectures for emerging warehouse-scale computing environments. In *Proc. International Symposium on Computer Architecture (ISCA)*, 2008. DOI: 10.1109/isca.2008.37. 5, 24

[102] K. Lim, Y. Turner, J. Santos, A. AuYoung, J. Chang, P. Ranganathan, and T. Wenisch. System-level implications for expansion and sharing in blade servers. In *Proc. International Symposium on High-Performance Computer Architecture (HPCA)*, 2012. 37

[103] D. Lo, L. Cheng, R. Govindaraju, L. Barroso, and C. Kozyrakis. Towards energy proportionality for large-scale latency-critical workloads. In *Proc. International Symposium on Computer Architecture*, 2014. DOI: 10.1109/isca.2014.6853237. 30

[104] P. Lotfi-Kamran, B. Grot, M. Ferdman, O. Kocberber S. Volos, J. Picorel, A. Adileh, S. Idgunji D. Jevdjic, E. Ozer, and B. Falsafi. Scale-out processors. In *Proc. International Symposium on Computer Architecture*, 2012. DOI: 10.1109/isca.2012.6237043. 6, 24

[105] Y. Low, D. Bickson, J. Gonzalez, C. Guestrin, A. Kyrola, and J. Hellerstein. Distributed GraphLab: A framework for machine learning and data mining in the cloud. In *Prpoc. VLDB Endowment*, 2012. DOI: 10.14778/2212351.2212354. 19, 20, 68

[106] Y. Low, J. Gonzalez, A. Kyrola, D. Bickson, C. Guestrin, and J. Hellerstein. GraphLab: pA new parallel framework for machine learning. In *Proc. Conference on Uncertainty in Artificial Intelligence (UAI)*, 2010. 19, 68

[107] K. Malladi, F. Nothaft, K. Periyathambi, B. Lee, C. Kozyrakis, and M. Horowitz. Towards energy-proportional datacenter memory with mobile DRAM. In *Proc. International Symposium on Computer Architecture (ISCA)*, 2012. DOI: 10.1109/isca.2012.6237004. 32, 33, 34, 35, 39

[108] K. Malladi, I. Shaeffer, L. Gopalakrishnan, D. Lo, B. Lee, and M. Horowitz. Rethinking DRAM power modes for energy proportionality. In *Proc. International Symposium on Microarchitecture (MICRO)*, 2012. DOI: 10.1109/micro.2012.21. 35

[109] J. Mars and L. Tang. Whare-Map: Heterogeneity in homogeneous warehouse-scale computers. In *Proc. International Symposium on Computer Architecture*, 2013. DOI: 10.1145/2485922.2485975. 30

[110] J. Mars, L. Tang, R. Hundt, K. Skadron, and M. Soffa. Bubble-up: Increasing utilization in modern warehouse scale computers via sensible co-locations. In *Proc. International Symposium on Microarchitecture*, 2011. DOI: 10.1145/2155620.2155650. 45, 46

[111] M. Martin, D. Sorin, B. Beckmann, et al. Multifacet's general execution-driven multiprocessor simulator (GEMS) toolset. *SIGARCH Computer Architecture News*, 33(4), 2005. DOI: 10.1145/1105734.1105747. 56

[112] D. Meisner, J. Wu, and T. Wenisch. BigHouse: A simulation infrastructure for data center systems. In *Proc. International Symposium on Performance Analysis of Systems and Software (ISPASS)*, 2012. DOI: 10.1109/ispass.2012.6189204. 83

[113] Micron. Calculating memory system power for DDR3. *Technical Note TN-41-01*, 2007. 33

[114] J. Miller, H. Kasture, G. Kurian, C. Gruenwald, N. Beckmann, C. Ceilio, J. Eastep, and A. Agarwal. Graphite: A distributed parallel simulator for multicores. In *Proc. International Symposium on High-Performance Computer Architecture (HPCA)*, 2010. DOI: 10.1109/hpca.2010.5416635. 79, 84

[115] R. Nishtala, H. Fugal, S. Grimm, M. Kwiatkowski, H. Lee, H. Li, et al. Scaling Memcache at Facebook. In *Proc. Symposium on Networked Systems Design and Implementation*, 2013. 12, 13, 40

[116] S. Novakovic, A. Daglis, E. Bugnion, B. Falsafi, and B. Grot. Scale-out NUMA. In *Proc. International Conference on Architectural Support for Programming Languages and Operating Systems (ASPLOS)*, 2014. DOI: 10.1145/2541940.2541965. 37, 40

[117] D. Ongaro, S. Rumble, R. Stutsman, J. Ousterhout, and M. Rosenblum. Fast crash recovery in RAMCloud. In *Proc. International Symposium on Operating Systems Principles (SOSP)*, 2011. DOI: 10.1145/2043556.2043560. 14

[118] Open Source. Memcached. `http://memcached.org/`. Accessed: 2015-02-03. 13

[119] Open Source. Oprofile. `http://oprofile.sourceforge.net`. Accessed: 2015-09-27. 43

[120] Open Source. Project Voldemort. `www.project-voldemort.com/`. Accessed: 2015-02-03. 13

[121] Open Source. QEMU: open source processor emulator. `http://wiki.qemu.org/Main_Page`. Accessed: 2015-06-09. 57

[122] J. Ousterhout, P. Agrawal, D. Erickson, C. Kozyarkis, J. Leverich, D. Mazieres, et al. The case for RAMCloud. *Communications of the ACM (CACM)*, 2011. DOI: 10.1145/1965724.1965751. 12, 13, 14, 37, 40

[123] K. Ousterhout, P. Wendell, M. Zaharia, and I. Stoica. Sparrow: Distributed, low latency scheduling. In *Proc. Symposium on Operating System Principles (SOSP)*, 2013. DOI: 10.1145/2517349.2522716. 78

[124] V. Pandey, W. Jiang, Y. Zhou, and R. Bianchini. DMA-aware memory energy management. In *Proc. International Sympsoium on High Performance Computer Architecture (HPCA)*, 2006. DOI: 10.1109/hpca.2006.1598120. 34

[125] D. Parkes, A. Procaccia, and N. Shah. Beyond dominant resource fairness: Extensions, limitations and indivisibilities. In *Proc. Conference on Electronic Commerce (EC)*, 2012. DOI: 10.1145/2229012.2229075. 50

[126] A. Patel, F. Afram, S. Chen, and K. Ghose. MARSSx86: A full system simulator for x86 CPUs. In *Design Automation Conference (DAC)*, 2011. 25, 56

[127] A. Patel, F. Afram, B. Fitzgerald, et al. MARSSx86 - Microarchitectural and system simulator for x86-based systems. `http://marss86.org/~marss86/index.php/Home`. Accessed: 2015-06-09. 58

[128] A. Patel, F. Afram, B. Fitzgerald, et al. Tutorial for MARSS: Microarchitectural systems simulator. In *Proc. International Symposium on Microarchitecture (MICRO)*, 2012. 57

[129] S. Pelley, P. Chen, and T. Wenisch. Memory persistency. In *Proc. International Symposium on Computer Architecture (ISCA)*, 2009. DOI: 10.1109/isca.2014.6853222. 37

[130] E. Perelman, G. Hamerly, M. Van Biesbrouck, T. Sherwood, and B. Calder. Using SimPoint for accurate and efficient simulation. In *Proc. International Conference on Measurement and Modeling of Computer Systems*, 2003. DOI: 10.1145/781027.781076. 78

[131] A. Putnam, A. Caulfield, E. Chung, D. Chiou, K. Constantinides, J. Demme, et al. A reconfigurable fabric for accelerating large-scale datacenter services. In *Proc. International Symposium on Computer Architecture (ISCA)*, 2014. DOI: 10.1109/isca.2014.6853195. 31

[132] M. Qureshi, A. Jaleel, Y. Patt, S. Steely, and J. Emer. Adaptive insertion policies for high-performance caching. In *Proc. International Symposium on Computer Architecture (ISCA)*, 2007. DOI: 10.1145/1250662.1250709. 32

[133] M. Qureshi, V. Srinivasan, and J. Rivers. Scalable high performance main memory systems using phase change memory. In *Proc. International Symposium on Computer Architecture (ISCA)*, 2009. DOI: 10.1145/1555754.1555760. 37, 56

[134] A. Raghavan, L. Emurian, L. Shao, M. Papaefthymiou, K. Pipe, T. Wenisch, and M. Martin. Computational sprinting on a hardware/software testbed. In *Proc. International Conference on Architectural Support for Programming Languages and Operating Systems (ASPLOS)*, 2013. DOI: 10.1145/2451116.2451135. 30

[135] A. Raghavan, Y. Luo, A. Chandawalla, M. Papefthymiou, K. Pipe, T. Wenisch, and M. Martin. Computational sprinting. In *Proc. International Symposium on High Performance Computer Architecture (HPCA)*, 2012. DOI: 10.1109/hpca.2012.6169031. 30

[136] C. Ranger, R. Raghuraman, A. Penmetsa, G. Bradski, and C. Kozyrakis. Evaluating MapReduce for multi-core and multiprocessor systems. In *Proc. International Symposium on High Performance Computer Architecture (HPCA)*, 2007. DOI: 10.1109/hpca.2007.346181. 18

[137] V.J. Reddi, B.C. Lee, T. Chilimbi, and K. Vaid. Web search using mobile cores: Quantifying and mitigating the price of efficiency. In *Proc. International Symposium on Computer Architecture (ISCA)*, 2010. DOI: 10.1145/1815961.1816002. 7, 8, 24, 25, 26, 27

[138] V.J. Reddi, B.C. Lee, T. Chilimbi, and K. Vaid. Mobile processors for energy-efficient web search. *Transactions on Computer Systems (TOCS)*, 2011. DOI: 10.1145/2003690.2003693. 7, 8, 24, 25, 26, 27

[139] G. Ren, E. Tune, T. Moseley, Y. Shi, S. Rus, and R. Hundt. Google-wide profiling: A continuous profiling infrastructure for data centers. *IEEE Micro*, 2010. DOI: 10.1109/mm.2010.68. 43

[140] P. Rosenfield, E. Cooper-Balis, and B. Jacob. DRAMSim2: A cycle accurate memory system simulator. In *Computer Architecture Letters (CAL)*, 2011. DOI: 10.1109/l-ca.2011.4. 25, 56, 57, 59, 61

[141] S. Rumble, D. Ongaro, R. Stutsman, M. Rosenblum, and J. Ousterhout. It's time for low latency. In *Proc. Conference on Hot Topics in Operating Systems (HotOS)*, 2011. 40

[142] N. Satish, N. Sundarama, M. Patwary, J. Seo, J. Park, M. Hassaan, S. Sengupta, Z. Yin, and P. Dubey. Navigating the maze of graph analytics frameworks using massive graph datasets. In *Proc. International Conference on Management of Data (SIGMOD)*, 2014. DOI: 10.1145/2588555.2610518. 20, 38

[143] J. Seo, J.L Park, J. Shin, and M. Lam. Distributed sociaLite: A datalog-based language for large-scale graph analysis. In *Proc. VLDB Endowment*, 2013. DOI: 10.14778/2556549.2556572. 20

[144] Y. Shao, B. Reagen, G. Wei, and D. Brooks. Aladdin: A pre-RTL, power-performance accelerator simulator enabling large design space exploration of customized architectures. In *Proc. International Symposium on Computer Architecture (ISCA)*, 2014. DOI: 10.1109/isca.2014.6853196. 56

[145] T. Sherwood, E. Perelman, and B. Calder. Basic block distribution analysis to find periodic behavior and simulation points in applications. In *Proc. International Conference on Parallel Architectures and Compilation Techniques (PACT)*, 2001. DOI: 10.1109/pact.2001.953283. 78

[146] T. Sherwood, E. Perelman, G. Hamerly, and B. Calder. Automatically characterizing large scale program behavior. In *Proc. International Conference on Architectural Support for Programming Languages and Operating Systems (ASPLOS)*, 2002. DOI: 10.1145/605397.605403. 78

[147] C. Silverstein, M. Henzinger, H. Marais, and M. Moricz. Analysis of a very large web search engine query log. In *Proc. Special Interest Group on Information Retrieval (SIGIR) Forum*, 1999. DOI: 10.1145/331403.331405. 8, 9

[148] Standard Performance Evaluation Corporation. SPEC CPU 2006. `https://www.spec.org/cpu2006/`. Accessed: 2015-02-03. 22

[149] Standard Performance Evaluation Corporation. SPEC SFS 2014. `https://www.spec.org/sfs2014/`. Accessed: 2015-02-03. 21

[150] Standard Performance Evaluation Corporation. SPECjbb 2010. `https://www.spec.org/jbb2015/`. Accessed: 2015-12-31. 21

[151] Standard Performance Evaluation Corporation. SPECjEnterprise 2010. `https://www.spec.org/jEnterprise2010/`. Accessed: 2015-02-03. 21

[152] I. Sutherland. A futures market in computer time. *Communications of the ACM*, 1968. DOI: 10.1145/363347.363396. 47

[153] J. Talbot, R. Yoo, and C. Kozyrakis. Phoenix++: Modular MapReduce for shared-memory systems. In *Proc. International Workshop on MapReduce and its Applications*, 2011. DOI: 10.1145/1996092.1996095. 18

[154] J. Teevan, E. Adar, R. Jones, and M. Potts. Information re-retrieval: Repeat queries in Yahoo's logs. In *Proc. Special Interest Group on Information Retrieval (SIGIR) Forum*, 2007. DOI: 10.1145/1277741.1277770. 7

[155] Transaction Processing Performance Council. TPC Benchmarks. `http://www.tpc.org/information/benchmarks.asp`. Accessed: 2015-02-03. 21

[156] K. Trivedi. *Probability and Statistics with Reliability, Queueing and Computer Science Applications*. Wiley, 2002. 80, 81

[157] K. VanCraeynest, S. Akram, W. Heirman, A. Jaleel, and L. Eeckhout. Fairness-aware scheduling on single-isa heterogeneous multi-cores. In *Proc. International Conference on Parallel Architectures and Compilation Techniques*, 2013. 49

[158] K. VanCraeynest, A. Jaleel, L. Eeckhout, P. Narvaez, and J. Emer. Scheduling heterogeneous multi-cores through performance impact estimation (pie). In *Proc. International Symposium on Computer Architecture*, 2012. DOI: 10.1109/isca.2012.6237019. 49

[159] L. Wang, J. Zhan, C. Luo, Y. Zhu, Q. Yang, Y. He, et al. BigDataBench: A big data benchmark suite from internet services. In *Proc. International Symposium on High Performance Computer Architecture (HPCA)*, 2014. DOI: 10.1109/hpca.2014.6835958. 6

[160] Q. Wang and B. Lee. Modeling communication costs in blade servers. In *Proc. Workshop on Power-Aware Computing and Systems (HotPower)*, 2015. DOI: 10.1145/2818613.2818743. 37, 48

[161] T. Wenisch, R. Wunderlich, B. Falsafi, and J. Hoe. TurboSMARTS: Accurate microarchitecture simulation sampling in minutes. In *Proc. International Conference on Measurement and Modeling of Computer Systems*, 2005. DOI: 10.1145/1064212.1064278. 78

[162] D. Wong and M. Annavaram. KnightShift: Scaling the energy proportionality wall through server-level heterogeneity. In *Proc. International Symposium on Microarchitecture (MICRO)*, 2012. DOI: 10.1109/micro.2012.20. 29

[163] S. Woo, M. Ohara, E. Torrie, J. Singh, and A. Gupta. The SPLASH-2 programs: Characterization and methodological considerations. In *Proc. International Symposium on Computer Architecture (ISCA)*, 1995. DOI: 10.1145/223982.223990. 22

[164] L. Wu, R. Barker, M. Kim, and K. Ross. Navigating big data with high-throughput, energy-efficient data partitioning. In *Proc. International Symposium on Computer Architecture (ISCA)*, 2013. DOI: 10.1145/2485922.2485944. 31

[165] L. Wu, A. Lottarini, T. Paine, M. Kim, and K. Ross. Q100: The architecture and design of a database processing unit. In *Proc. International Conference on Architectural Support for Programming Languages and Operating Systems (ASPLOS)*, 2014. DOI: 10.1145/2541940.2541961. 31

[166] W. Wu and B. Lee. Inferred models for dynamic and sparse hardware-software spaces. In *Proc. International Symposium on Microarchitecture*, 2012. DOI: 10.1109/micro.2012.45. 43, 44

[167] R. Xin, J. Gonzalez, M. Franklin, and I. Stoica. GraphX: A resilient distributed graph system on Spark. In *Proc. International Workshop on Graph Data Management Experience and Systems (GRADES)*, 2013. DOI: 10.1145/2484425.2484427. 20

[168] H. Yang, A. Breslow, J. Mars, and L. Tang. Bubble-flux: Precise online QoS management for increased utilization in warehouse scale computers. In *Proc. International Symposium on Computer Architecture*, 2013. DOI: 10.1145/2485922.2485974. 45, 46

[169] M. Youst. PTLsim: A cycle accurate full system x86-64 microarchitectural simulator. In *Proc. International Symposium on Performance Analysis of Systems and Software*, 2007. DOI: 10.1109/ISPASS.2007.363733. 57

[170] M. Zaharia, A. Konwinski, A. Joseph, R. Katz, and I. Stoica. Improving MapReduce performance in heterogeneous environments. In *Proc. Conference on Operating Systems Design and Principles (OSDI)*, 2008. 49

[171] M. Zaharia. Slides: Introduction to Spark internals. `https://cwiki.apache.org/confluence/display/SPARK/Spark+Internals`. Accessed: 2015-06-09. 65, 78, 82

[172] M. Zaharia, M. Chowdhury, T. Das, A. Dave, J. Ma, M. McCauley, M. Franklin, S. Shenker, and I. Stoica. Resilient distributed datasets: A fault-tolerant abstraction for in-memory clustering computing. In *Proc. Symposium on Networked Systems Design and Implementation*, 2012. 64

[173] M. Zaharia, M. Chowdhury, M. Franklin, S. Shenker, and I. Stoica. Spark: Cluster computing with working sets. In *Proc. HotCloud*, 2010. 64

[174] S. Zahedi and B. Lee. REF: Resource elasticity fairness with sharing incentives for multiprocessors. In *Proc. International Conference on Architectural Support for Programming Languages and Operating Systems (ASPLOS)*, 2014. DOI: 10.1145/2541940.2541962. 51, 52, 53

[175] S. Zahedi and B. Lee. Sharing incentives and fair division for multiprocessors. *IEEE Micro*, 2015. DOI: 10.1109/mm.2015.49. 51, 52, 53

[176] Y. Zhang, B. Jansen, and A. Spink. Time series analysis of a web search engine transaction log. *Information Processing and Management*, 2009. DOI: 10.1016/j.ipm.2008.07.003. 10

[177] P. Zhou, B. Zhao, J. Yang, and Y. Zhang. A durable and energy efficient main memory using phase change memory. In *Proc. International Symposium on Computer Architecture (ISCA)*, 2009. DOI: 10.1145/1555754.1555759. 37, 56

Author's Biography

BENJAMIN C. LEE

Benjamin C. Lee is the Nortel Networks Associate Professor of Electrical Engineering and Computer Science at Duke University. His research focuses on power-efficient architectures and emerging technologies for high-performance computer systems. He is also interested in the economics and public policy of computation. At Berkeley, he developed auto-tuning frameworks for sparse linear algebra. At Harvard, he introduced statistical machine learning for processor design. At Duke, Dr. Lee leads the System Integration Architecture Laboratory, which studies datacenter design and management with economic mechanisms and game theory.

Dr. Lee received his B.S. in Electrical Engineering and Computer Science from the University of California at Berkeley, his Ph.D. in Computer Science at Harvard University, and his post-doctorate in Electrical Engineering at Stanford University. He has held visiting positions at Microsoft Research, Intel Labs, and Lawrence Livermore National Lab. Dr. Lee has received the NSF CAREER Award and the NSF Computing Innovation Fellowship. His research has been honored twice as Top Picks by *IEEE Micro Magazine* and twice as Research Highlights by the *Communications of the ACM*.